Advances in Bioclimatology 5

Springer

Berlin
Heidelberg
New York
Barcelona
Budapest
Hong Kong
London
Milan
Paris
Santa Clara
Singapure
Tokyo

Volumes in the series

Vol. 1: R. L. Desjardins, R. M. Gifford, T. Nilson, E. A. N. Greenwood (1992)

Vol. 2: J. D. Kalma, G. P. Laughlin, J. M. Caprio, P. J. C. Hamer
The Bioclimatology of Frost.
Its Occurrence, Impact and Protection (1992)

Vol. 3: Y. Cohen, J. M. Elwood, M. G. Holmes, V. A. Kanevski, S. M. Kochubey, J. Ross, T. Shadchina, E. Simensen, F. I. Woodward (1994)

Vol. 4: D. J. Beerling, P. W. Brown, K. G. Hubbard, J. D. Kalma, S. J. Meyer, M. Nunez, O. D. Sirotenko, R. L. Snyder, G. Stange, F. J. Woodward (1996)

Vol. 5: A. Auliciems, R. de Dear, M. Fagence, L. S. Kalkstein, S. Kevan, S. V. Szokolay, A. R. Webb (1997)

Advances in Bioclimatology_5

Human Bioclimatology

With Contributions by
A. Auliciems R. de Dear M. Fagence
L. S. Kalkstein S. Kevan S. V. Szokolay
A. R. Webb

With 33 Figures

Springer

Edited by
Andris Auliciems
The University of Queensland
Department of Geographical Science
and Planning
Brisbane, Queensland 4072
Australia

ISBN-13:978-3-642-80421-2 e-ISBN-13:978-3-642-80419-9
DOI:10.1007/978-3-642-80419-9

Cataloging - in - Publication Data applied for

Die Deutsche Bibliothek - CIP-Einheitsaufnahme
Human bioclimatology / ed: Andris Auliciems. - Berlin ; Heidelberg ; New York ; London ; Paris ; Tokyo ; Hong Kong ; Barcelona ; Budapest : Springer, 1997
(Advances in bioclimatology ; Vol. 5)

Softcover reprint of the hardcover 1st edition 1998

Production: PRO EDIT GmbH, 69126 Heidelberg, Germany
Typesetting: Storch GmbH, 97353 Wiesentheid, Germany
SPIN: 10541121 31/3137 - 5 4 3 2 1 0 - Printed on acid-free paper

Preface to the Series

Advances in Bioclimatology – the study of the relations between the physical environment and the form and function of living organisms – have been spectacular during the last third of this century.

Before this period, the subject, having slowly emerged from its classical origins as a branch of natural history, had reached the stage of a collection of largely empirical, statistical relationships between standardized but often inappropriate climatological and biological measurements.

Since then, research into the basic physical and physiological mechanisms involved has used the latest techniques of measurement and analysis to develop various bioclimatic relations which have contributed much to improving crop and animal production and optimizing the human environment.

Recently, some of these relationships have been incorporated into larger models of climate-ecosystem interactions. Such models are being used to assess the often unintended effects of human activity on various elements of the biosphere.

However, the advances described have been very unevenly spread through the vast field of interest encompassed by bioclimatology; the fields of plant, animal and human climatology have largely advanced in independent fashions and even within each biological province different techniques of analysis and measurement have developed for different time and space scales of organization.

One of the major aims of this new review series is to overcome this separate development by providing a common forum for those wishing to obtain an authoritative review of the latest developments in bioclimatology.

The emphasis will be on advances which are soundly based on biological and physical principles rather than those describing empirical relationships. Reviews will also deal with the latest techniques of measurement and analysis which are of relevance to bioclimatology and to those describing broader ecological studies in which bioclimatology provides a major element.

Although most of the reviews to be published will be commissioned, the editors would welcome suggestions from individuals interested in contributing a review of the type described, as well as for ideas on major topics of wide interest around which a number of individual reviews could be centered.

Bet Dagan, Israel

G. Stanhill
Editor

List of Editors

Managing Editor

Those interested in contributing a review to this series are invited to contact Dr. Gerald Stanhill or one of the Associate Editors. Proposals should outline briefly the review's aims and scope.

Contents

Human Bioclimatology: An Introduction

A. AULICIEMS

Living organisms respond to atmospheric variability and variation, and over time morphological and process differentiations occur both within individuals and the species, as well as in the environment itself. In systems language, the concern is with the atmospheric process-response system of energy and matter flows within the biosphere. The study of such interactions between living organisms and the atmospheric environment falls within the field of bioclimatology, alternatively referred to as biometeorology.

Amongst the more readily recognizable study areas under the bioclimatological umbrella are those that investigate the effects of atmospheric variation and variability upon

1. Terrestrial and aquatic ecology (zoological, botanical and ethological), natural resource production and management (including silviculture, agriculture, horticulture, and grassland, wetland, and marine systems).
2. Stress, morbidity and mortality in animals and humans (including physiological and psychological adaptations).
3. The built environment (all aspects of planning, urban design, and architecture).
4. Economic systems and social activities (including organizational, individual, and group behavior and management).

In addition, bioclimatology is very much concerned with the feedback loop, that is both

5. The inadvertent modification of the atmosphere by living systems, especially human, i.e., studies of pollution, changes to atmospheric amenity, and the processes of deterioration of landscape (deforestation and desertification), and
6. The advertent modifications of natural energy and matter flows within urban areas and indoor climate constructions.

It becomes evident that bioclimatological studies are not constrained by physical or temporal scales, nor by the strength of the atmospheric signal, which can range from minor random variability to the extremes of geophysical events: the hydrometeor, radiation, and thermal hazards. Clearly, within such broad boundaries, bioclimatology is both ubiquitous to traditional disciplinary areas as well as being quintessentially interdiscliplinary.

Traditionally bioclimatology has been particularly concerned with atmospheric impacts upon humans and their technocultural creations, and especially with the growing awareness of the human factor in the modificiation of environmental and the atmospheric processes themselves. Nor is this anthropocentric view inapparopriate, since in the modern holistic paradigm, ecological concern without reference to human-generated atmospheric process modifications, or indeed to the resulting impacts on humans themselves, would be unrealistic.

Bioclimatology is about change, and is regarded on its biological side as being a science of adaptation (Weihe 1994). Where human adaptations are concerned, several well-established principles appear to govern the adaptation processes. The first and most obvious is that responses are not constant but vary in accordance with stimulus duration, frequency, and intensity. The second may be related to the Weber-Fechner stimulus-response model (Stevens and Stevens 1963), within which psycho-physiological responses will occur over the whole range of stimuli and the amount of stress experienced is genereally proportional to the log of the signal strength until at high intensities, the capacity for further response ceases. The third is that the amount of stress experienced decreases with acclimatization, which is both physiological and psychological (Sargent 1963). The fourth is that the relative position of any "optimum" is not constant, but shifts in the direction of the exposure (Helson 1964; Wohlwill 1974), as for example with thermal acclimatization. The fifth is that repeated exposure to a particular stimulus promotes habituation and in turn a narrowing of the general tolerance band (Weihe and Mertens 1991). The sixth is that involuntary exposure becomes particularly stressful (Starr 1969), and the seventh, stated most simply, is that the continuous capacity for response has limits (Selye 1957).

With regard to study paradigms, human adaptation can seemingly be examined within three overlapping categories of interactions with the atmosphere. The first and dominant of these may be summarized as the homeostatic, or the biologically steady-state responses that, once unsuccessful, may incur decrement in performance and increased morbidity and even mortality. The second category is atmospheric hazard and risk impacts and adjustments, being cultural and individual preference-dependent adaptations and adjustments to relatively infrequent but high-intensity atmospheric events. The third area of concern is with perception, behavior, and planning of the atmospheric environment as an amenity and/or resource, usually being a function of preferred scheduling, location, or economic choice. These three main categories may be visualized to lie on a highly generalized continuum as illustrated in Fig. 1.

In a systems sense, the links between the main components are more formally represented in Fig. 2. This is based on the drought model of Warwick (1980) and discussions by Kates (1985), but emphasizes the otherwise neglected first-order impact and incorporates principal feedbacks and the all-important perception-choice control that is central to natural hazards research (Burton et al. 1978; Whyte 1985). An illustration of this model can be a simple example: a cold

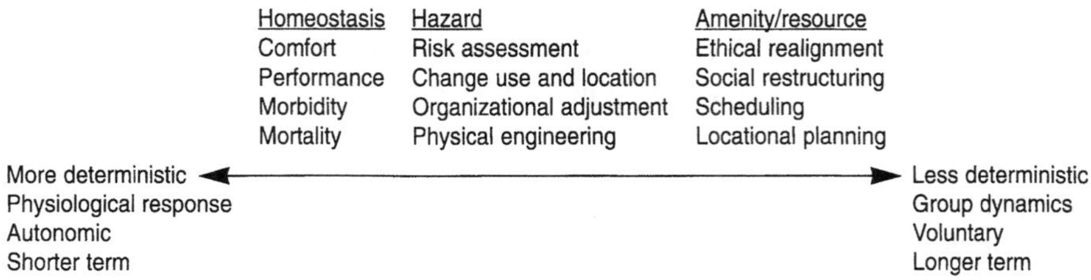

Fig. 1. Generalized continuum of human responses to atmospheric stimulus

winter may in general tend to lower indoor temperatures and increase the frequency of cold discomfort (first-order impact). This is likely to promote a greater demand for more space heating and thus, if so preferred by people, for fossil burning (second-order impact), which in turn, however, may decrease the availability of funds for alternative purposes (and technologies) and in general lower the overall well-being of the whole society (third-order impacts). In the short term, at least, technology and socioeconomic arrangements enable increased space heating and thereby decrease cold discomfort (feedbacks loops *a* and *b*), but there is increased generation of radiative gases (feedback loop *c*), which in the long term may contribute to global warming, but also, ultimately, to warmer winters that may actually tend to reduce the cold stimulus. Human technological and socioeconomic responses, however, are to a large degree controlled by culturally based attitudes, perceptions, and evaluations (feedback *e*) of the stimuli and the alternatives for responding.

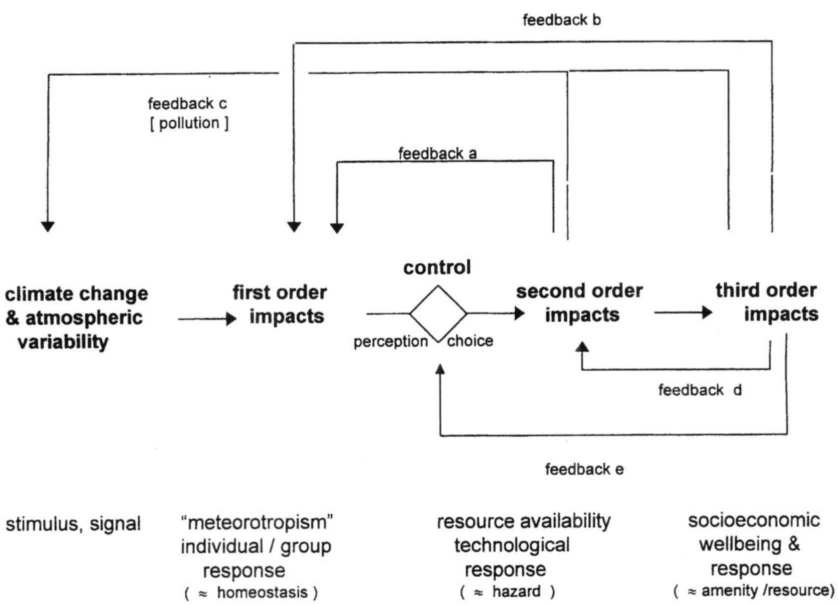

Fig. 2. An impacts model of human bioclimatology

Clearly, human-atmosphere interrelationships are not simple, and the well-being of people and the condition of the atmospheric environment, both indoors and outdoors, are inseparable and need to be viewed as holistically as possible. Increasingly, there is a recognition that deterministic human responses, as at times assumed in some engineering solutions, occur mainly at the first level of impacts, and that human perceptions and choices are the basis for controls, adaptations, and, at times, environmental inclemency.

The six reviews in this volume discuss particulars at the various impact levels, but inevitably the topics selected are only some of those possible. The first, by Ann Webb, deals with one of the most important of the atmospheric stimuli, solar ultraviolet radiation. Recent developments in the measurement of ultraviolet radiation are reviewed and their first-order impacts upon living organisms – humans, animals, plants, and aquatic life – are summarized. Increases in ultra-shortwave radiation due to anthropocentric generation of radiative gases must pose one of the most profound problems facing humankind over the next century.

The attainment of thermal comfort through indoor climate modification has been hailed by Benzinger (1978) as one of humanity's great achievements. However, the second review, on thermal adaptation and variable indoor climate control, contributed by Andris Auliciems and Richard de Dear, presents the other side of the coin, demonstrating that microclimate management approaches in technologically advanced countries over the past half-century may go actually go against several of the general principles of adaptation as outlined above. A case is presented for new management strategies for maximization of health and comfort and minimization of external energy usage, and therefore a reduction in radiative gas releases.

This second-order-impacts category is further examined in Richard de Dear's review of the modern-day colorfully named "sick building syndrome". This largely represents the second-order impacts cateogry, as resulting from ill-applied technology initially put in place to reduce the direct impacts of weather and climate upon daily living in the built environment. As in most other cases, however, the impacts are not simple, and elements of third order can be seen to play as significant a role as the atmospheric stimulus itself.

The theme of rationalization of indoor climate design is continued by Steve Szokolay in his review of bioclimatic architecture and solar energy. Here, the beneficial dimensions of short-wave radiation is examined from the viewpoint of technological design as related to the human requirements touched upon in the preceding reviews.

Human mechanisms of adaptation to atmospheric variability may range from basic short-term and autonomic physiological responses to complicated behavioral adjustments. The avoidance response to adverse stimulus and its antithesis, the atmospheric amenity search, is discussed by Mike Fagence and Simon Kevan in the fifth review. Methods of quantification and assessment of atmospheric amenity are discussed, and illustration is provided of the complexities of analyzing an area that abounds with first-, second-, and third-order impacts.

The firmest indications of adverse impacts are manifested in human morbidity and mortality itself. Larry Kalkstein's review, "Climate and Human Mortality: Relationships and Mitigating Measures", analyzes the incidence of mortality in terms of prevailing climatic conditions and speculates on the relationships under assumptions of global warming. Heat-stress-related deaths are seen to increase but the author takes a pragmatic look at the possibilities of reducing the severity of the impacts.

Clearly, in a field as vast as that of human bioclimatology, each of the above topics raises possibilities of other interpretations and emphasis. In addition, there are areas not covered in even a cursory way, especially first- and higher-order impacts relating to human performance and behavior, which lie beyond simple avoidance or amenity responses. Particularly as related to the potential for global climate change, there is an urgent need for current reviews and syntheses relating climate and such phenomena as:

- Risk taking, decision making, and management
- Regional and town planning
- Dietary behavior
- Alcohol and other drug consumption
- Violence, crime, and other antisocial manifestations
- Work performance, absenteeism, and accidents
- Use of public facilities: recreational, educational, and sport
- Retail and commerce
- Economic impacts and responses
- Military implications and applications

Innate abilities for adaptation have enabled humans to thrive in climates well removed from those of their evolutionary origins. Their technological and cultural adjustments have also enabled humans to cope effectively with a wide range of the shorter-term hydrometeor impacts, and otpimization of these facilities becomes a vital concern for all societies. Some of this is known and applied, yet, there is much to be researched in regard to the reduction of impacts upon the biosphere, and upon humanity itself, before we can satisfactorily define universally optimum climates or methods of coping. The interrelationships are many and complicated, and are subject to differences within the landscape and adaptations amongst the people exposed to the atmospheric stimulus. Thus, although the capacity for impact reduction at all levels of stimulus intensity and economic gain is already considerable, these differences make specific case studies a prerequisite for optimizing impact-mitigating measures. An overarching theory of human bioclimatology, or at least universally acceptable algorithms that adequately describe behavior and impacts, are a long way away.

Acknowledgements. Several individuals in particular deserve thanks for enabling this collection of reviews to be published. Firstly, Gerald Stanhill, Managing Editor of the Series, has been supportive and tolerant during some difficult moments. Jetse Kalma, also of the Editorial Board, has been a good friend and advisor. Following the premature passing away of Simon Kevan, his wife, Brenda, courageously gathered the material available and enabled the work to proceed. Many thanks are also due to the anonymous reviewers for their important inputs.

References

Auliciems A (1996) Integrating models of biometeorology. Paper presented to the Plenary Session of the 14th Congr Int Soc Biometeorology, Llubljana, 1–8 September 1996

Benzinger TH (1978) Introduction to the 1st Int Indoor Climate Symp at Copenhaben, 30 August–1 September 1978

Burton I, Kates RW, White G (1978) The environment as hazard. Oxford University Press, New York

Helson H (1964) Adaptation-level theory. Harper & Row, New York

Kates RW (1985) The interaction of climate and society. In: Kates RW, Ausubel J, Berberian M (eds) Climate impact and assessment, SCOPE 27. Wiley, New York, pp 3–36

Sargent F II (1963) Tropical neurasthenia: giant or windmill? In: Arid zone research. Environmental psychology and physiology, Proc Lucknow Symp, UNESCO, Paris

Selye H (1957) The stress of life. Longman, London

Starr C (1969) Social benefit versus technological risk. Science 165: 1232–1238

Stevens JC, Sevens SS (1963) The dynamics of subjective warmth and cold. In: Hardy JD (ed) Temperature, its measurement in science and industry, vol. 3. Reinhold, New York, pp 239–243

Warwick RA (1980) Drought in the Great Plains: a case study of research on climate and society in the USA. In: Ausubel J, Biswas J (eds) Climatic constraints and human activities. Pergamon, New York, pp 93–1132

Weihe WH (1994) The role of biometeorology in society. Presidential address. In: Maarouf AR, Barthakur NN, Haufe WO (eds) Proc 13th Int Congr of Biometeorology, part 2, vol 1, 12–18 September 1993, Calgary, pp 9–13

Weihe WH, Mertens R (1991) Human well being, disease and climate. In: Jager J, Ferguson HL (eds) Climate change: science, impacts and policy, Proc 2nd World Climate Conf, Cambridge University Press, Cambridge, pp 345–360

Whyte AVT (1985) Perception. In: Kates RW, Ausubel J, Berberian M (eds) Climate impact and assessment, SCOPE 27. Wiley, New York, pp 403–436

Wohlwill JF (1974) Human adaptation to levels of environment stimulation. Hum Ecol 2: 127–147

Techniques for Measuring Solar Ultraviolet Radiation

A. R. Webb[1]

1 Introduction

The Latin for "beyond" is ultra, thus ultraviolet (UV) radiation is that part of the electromagnetic spectrum having wavelengths beyond (shorter than) the violet end of the of the visible spectrum. The complete UV waveband covers the wavelength range 200–400 nm, but it is subdivided into three regions labelled UVA, UVB and UVC. The boundaries between the three classes of UV radiation, determined by the definition of UVB radiation, vary throughout the scientific literature, with the short-wavelength boundary usually lying between 280 and 290 nm and the long-wavelength boundary between 315 and 320 nm. The Commission Internationale d'Eclairage (CIE) define UVB radiation as the waveband 280–315 nm, but in discussions of solar UV radiation and its effects the wavelengths between 280 and 320 nm are often quoted as UVB. For the purposes of this article a precise statement of the UVB waveband is not necessary. The atmosphere imposes a varying short-wavelength limit on solar UVB radiation reaching the earth's surface, which for much of the globe is greater than 290 nm. At the long-wavelength boundary no broadband instrument or biological action spectrum has a total and abrupt end of response in the UVB/UVA cross-over region, so another fuzzy demarcation is not a disadvantage but a practical occurrence. The references cited should contain the respective authors' definition of UVB radiation in the work concerned, but for collective discussion of UV issues in this chapter UVB radiation will be taken to include all wavelengths from 280 to 320 nm. A number of other terms which require careful definition are listed in the glossary in Appendix C.

1.1 A Global View of UV Radiation

Radiation emitted by the sun is similar to that of a black body at approximately 6000 K and the extra-terrestrial solar spectrum has a UV component that accounts for 9.3% of the total solar radiation. The solar spectrum reaching the earth's surface has been modified by the atmosphere through which the radiation must pass. The wavelength dependence of the attenuation is determined by

Department of Physics, University of Manchester, Institute of Science and Technology, P.O.Box 88, Sackville Street, Manchester, M60 1QD, UK the atmospheric constituents and scattering processes. The magnitude of the attenuation is a function of the radiation pathlength through the atmosphere and the amount of each attenuator along that pathlength. The pathlength is determined by the solar zenith angle (z), which is itself a function of latitude and time (of day and year). Pathlength is approximately equal to $(\cos z)^{-1}$; therefore, when the sun is directly overhead ($z = 0$, e.g. noon at the equator at the equinox), the pathlength and the potential for attenuation are smallest.

Atmospheric attenuation occurs in a number of ways. A consistent attenuator of solar radiation is Rayleigh scattering. This process applies at all wavelengths, but scattering is proportional to λ^{-4} so the UV is the most strongly scattered waveband. Forward and backward scattering are approximately equal, so at short wavelengths a proportionally greater amount of radiation is removed from the direct beam and scattered back towards space or reaches the ground by an indirect route (diffuse radiation).

Efficient absorption of the UVC wavelengths by molecular oxygen and ozone in the upper atmosphere removes this part of the extra-terrestrial solar spectrum before it reaches the earth's surface. As wavelenth increases ozone continues to play a major absorptive role, but its effectiveness decreases sharply across the UVB portion of the spectrum (Fig. 1a): the absorption at 320 nm is only 1% of that at 280 nm, and at 340 nm the absorption by ozone is negligible (Barton and Paltridge 1979). The declining absorptive powers of ozone are mirrored in the ground-level solar spectrum by a steeply increasing short wave cut-on through the UVB (Fig. 1a). The position of this spectral edge, the shortest detectable wavelength to reach the surface, is a function of pathlength (Fig. 1b) but has a further dependence on latitude and season as determinants of stratospheric ozone concentration. Ozone climatology (Fig. 2a) shows the latitudinal dependence and seasonal cycles of long-term mean ozone concentrations. On a short timescale there are considerable year-to-year and day-to-day fluctuations about this mean. Since the recognition of the Antarctic ozone hole and the role of chlorofluorocarbons in its formation [(Farman et al. 1985; Farmer et al. 1987), first proposed by Molina and Rowland (1974)], there has been much concern that the mean stratospheric ozone will be depleted over other areas of the globe (Fig. 2b), with a consequent increase in UVB radiation at the surface, and its attendant effects.

In addition to the widespread and persistent UVB attenuators described above, there are several localised, transient atmospheric effects and small cyclical solar influences on UV radiation reaching the ground. Low-level pollutants, including ozone, absorb UVB wavelengths in the boundary layer most effectively because at this level much of the UV radiation is diffuse, with an elongated pathlength and a greater probability of absorption (Bruhl and Crutzen 1989). Aerosols in both the troposphere and the stratosphere attenuate direct beam radiation, but increased scattering can lead to more diffuse radiation and therefore little change in global irradiances at the surface (McKenzie et al. 1993). There are however, situations where measurements indicate significant reductions of UV irradiances due to low-level pollutants (Seckmeyer et al.

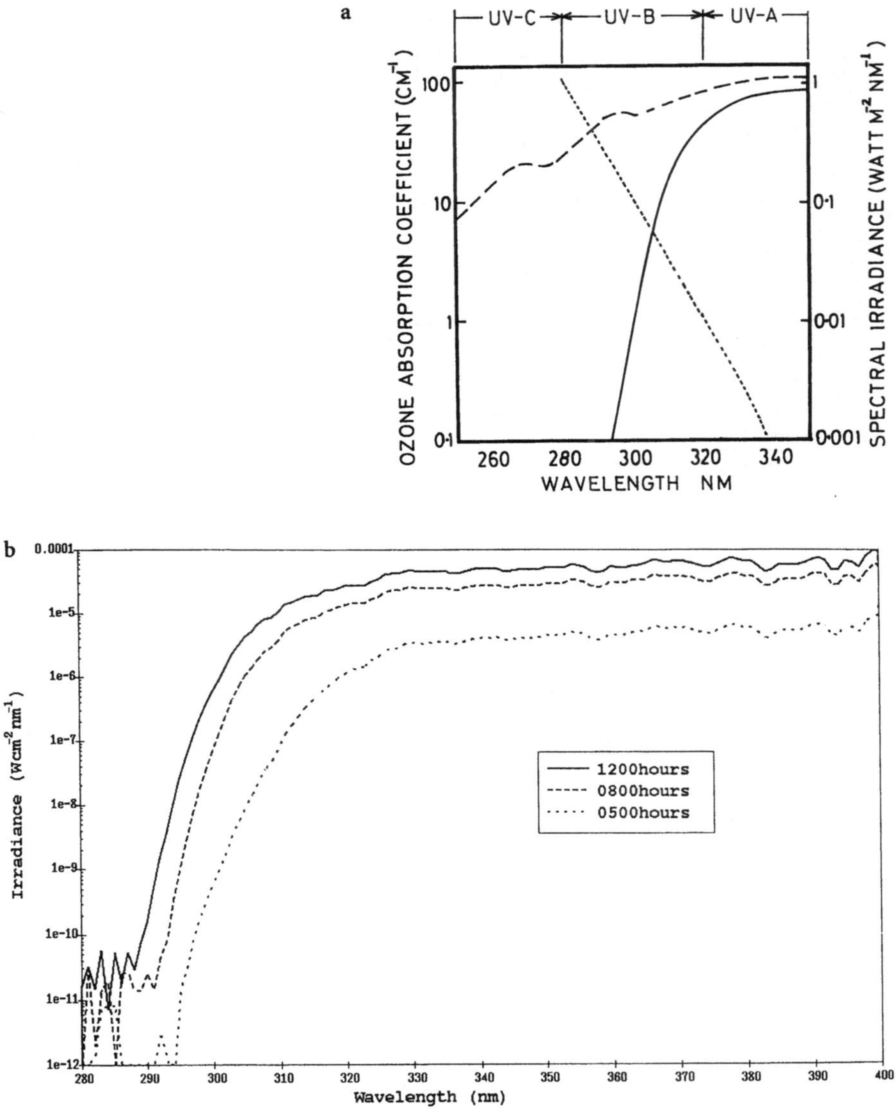

Fig. 1. a The extraterrestrial solar spectrum (-----), ozone absorption spectrum (......) and solar spectrum at the ground (——). **b** The solar spectrum measured at Reading, UK, on a clear day (14 June 1994) at 0500 ($z = 81°$), 0900 ($z = 53°$) and 1200 UT ($z = 28°$). As the zenith angle decreases, the limit of the solar spectrum shifts to shorter wavelengths. The measurements were made with an Optronic 742 spectroradiometer. Signals < 10^{-10} Wcm^{-2}nm^{-1} are electrical noise. Signals at the shortest detectable wavelengths are generally overestimates of the true irradiance because of the broad slit function (1.4 nm FWHM) of the instrument

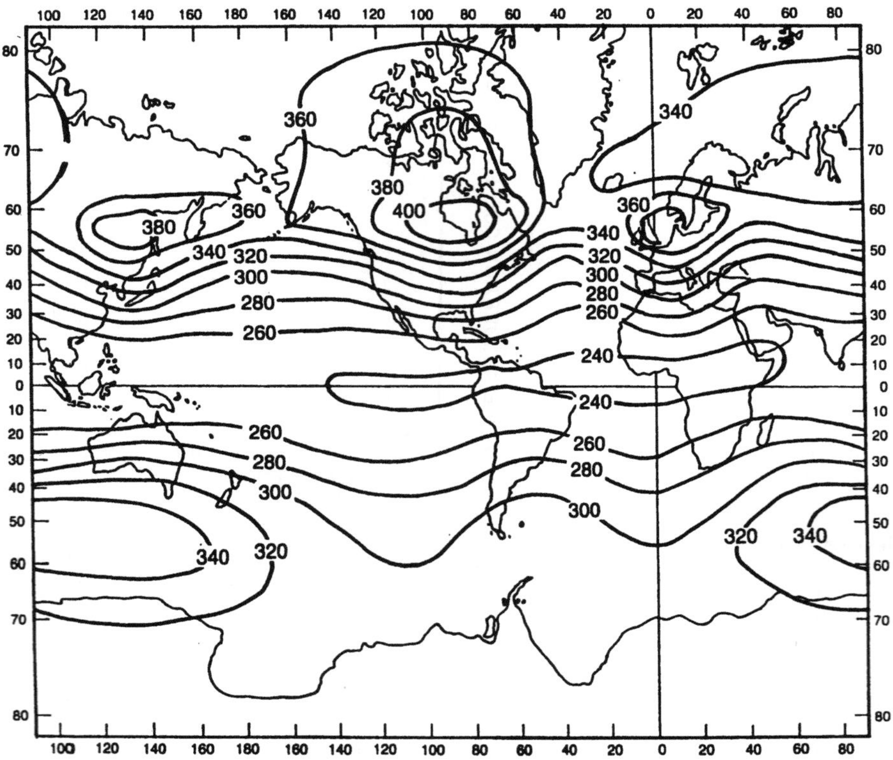

Fig. 2a. Average ozone climatology derived from ground-based measurements for the years 1957–1975 (WMO 1988)

1994), and model results also point to noticeable effects (Liu et al. 1991). Michelangeli et al. (1992) calculated that in certain conditions stratospheric aerosols could increase irradiances at short UVB wavelengths. Thus the overall effect of aerosols is variable and depends upon their type and distribution.

A more obvious attenuator is cloud. The degree to which cloud changes radiation incident at the surface will depend upon the type, thickness and spatial distribution of the cloud. For a layer of cloud attenuation increases with cloud depth. In broken cloud incident radiation can increase briefly due to forward scattering of circumsolar radiation from cloud edges, usually followed by a drop in irradiance as the cloud passes before the sun and blocks direct-beam radiation. This phenomenon occurs at all wavelengths. In contrast to ozone, cloud has no strong spectral selectivity in its absorption of UV radiation; it attenuates more equally throughout the UV and therefore does not change the shape of the spectrum greatly, merely reducing the intensity (Webb 1991; Bais et al. 1993). Evidence of some spectral influence of clouds is given by Seckmeyer et al.

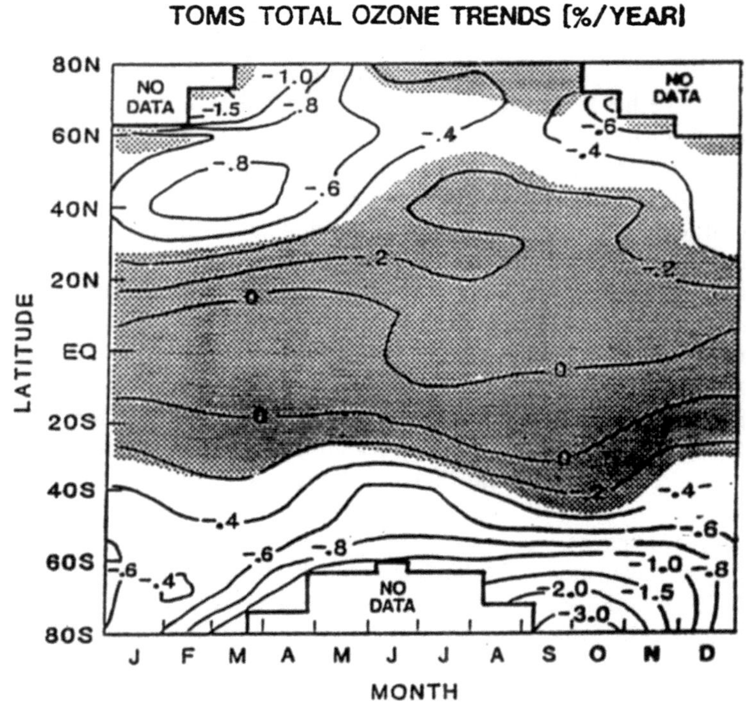

Fig. 2b. The percentage change per year in TOMS (Total Ozone Mapping Spectroradiometer) total ozone from 1978 to 1990 (Stolarski et al. 1991). The trend is not significant at the 2 σ level in the *shaded area*

(1994), but it is changes in ozone which have the greatest effect at the shortest wavelengths and therefore produce characteristic changes in spectral shape as well as overall intensity. Cloud is a feature of different weather patterns, and the dynamics of changing weather systems can have other indirect effects on UV radiation. High-pressure systems are associated with a high tropopause which pushes up into the stratosphere, squeezing stratospheric air out of this region and reducing column ozone. In the stratosphere over the corresponding low-pressure regions column ozone is increased. This process accounts for much of the day-to-day variation in ozone in mid-latitude regions where the passage of frontal systems is common.

Seasonal weather anomalies also add complexity to the study of climatology and trends. In 1992 and 1993 significantly low ozone was recorded over much of the globe (Gleason et al. 1993; Herman and Larko 1994; Komhyr et al. 1994), especially the populated regions of the northern hemisphere, where elevated UVB radiation was also measured (Kerr and McElroy 1993; Seckmeyer et al. 1994). While the elevated UVB showed the distinct spectral signature that associates it with low ozone, it should be noted that in these years there were unusually persistent regions of high pressure situated over Europe for much of the relevant winter and spring periods, and in addition stratospheric chemistry was still perturbed by the injection of aerosols from Mount Pinatubo in 1991 (SORG 1993). Both these events served to decrease ozone, and it remains to be seen whether these years were an extreme variation in ozone climatology or not. In 1994 ozone levels returned to values closer to the expected long-term trend (Scientific Assessment of Ozone Depletion: 1994). During the winter and spring of 1995 periods of low ozone were observed over the Arctic and high northern latitudes, and were closely monitored during the SESAME campaign. Results of many of the related SESAME activities, including UV measurements, will be reported in a special issue of Journal of Geophysical Research.

Finally, the extra-terrestrial solar spectrum is not a constant but changes on a number of time scales (Madronich 1993). The annual progression of the earth around the sun is slightly elliptical and the changing earth-sun distance results in extra-terrestrial intensities ±3.3% from the mean in January and July respectively. This does not change the UV climte at any one place, but it does result in a difference between northern and southern hemispheres, working with the seasons in the south and in opposition in the north. On a longer time scale, the 11-year sunspot cycle marks changes in solar output at the shortest extra-terrestrial wavelengths. The sunspot cycle has negligible effect on the solar output of UVB and UVA wavelengths (Madronich 1993), but changes at UVC wavelengths affect the chemical equlibrium of ozone production and destruction in the stratosphere, and by changing ozone it has an indirect effect on the transmission of UVB to the surface.

1.2 Determining UV Radition Intensity at the Ground

There are two ways in which UV radiation reaching the ground can be quantified; both methods have their advantages and their problems. One approach is to make direct measurements at a discrete number of locations: the topic of the rest of this review. With a well-characterised and well-calibrated instrument this will provide knowledge of the UV irradiance at specific locations at any time, but the worldwide coverage of UV monitoring stations is sparse (Webb and Weatherhead 1997), with large areas lacking any UV instrumentation.

1.2.1 UV Radiation Models

To achieve a global picture of UV irradiances radiative transfer models are required, but these in turn depend upon other measurements as input data, and can only ever be as good as the data supplied to the model. Various radiation schemes are available for calculating UV irradiances at the ground. Calculation of the direct-beam part of the irradiance is fairly straightforward given good knowledge of the attenuators in the atmosphere; assessing the diffuse radiation is more difficult and open to different approaches. Green et al. (1980) and

Schippnick and Green (1982) developed an empirical fit to determine diffuse radiation, a method that works well in the UV region for solar zenith angles less than 70°. Green's model has been extensively used, especially for climatology and ozone depletion effects studies. More recently, increased computing power has enabled general use of more complex analytical solutions to the diffuse radiative transfer equations; for example, the discrete-ordinate-method scheme of Stamnes et al. (1988) is widely used in the UV community.

To model solar UV iradiance at a given time and place as a direct comparison to a measurement, accurate knowledge is required of: (1) column ozone and its vertical distribution; (2) the vertical temperature profile; (3) aerosol amount, type and distribution; (4) other UV-absorbing compounds, e.g., SO_2; (5) areal surface albedo; and (6) cloud amount, type and distribution. Correct specification of the ozone is vital for calculating UVB irradiances but is irrelevant at the longer UVA wavelengths. Fortunately, good global coverage of total ozone amounts is available from a combination of satellite measurements and ground-based measurements, although ozone profile measurements are more sparse, and "typical" profiles are often used. Aerosol is usually incompletely defined and surface albedo has to be estimated. The most difficult parameter to model is cloud, although cloud estimates can be made from other meterological or satellite data, as can aerosol optical depth. Under clear sky conditions reasonably good agreement can be achieved between models and measurements (Forster 1994; Wang and Lenoble 1994), but difficulties of accurately specifying cloud inputs to the model make computations in overcast conditions less reliable. A major advantage of validated models is their ability to provide a global clear sky climatology of UV, using climatological, or temporally specific, values of ozone. This can then be used to assess the relative UV irradiances of areas where no measurements are available, and to predict changes in UV at the ground which may arise from changes in ozone, albedo or other model input parameters (Forster 1994; Madronich 1992). Potential changes in UV irradiances due to ozone depletion are often expressed in terms of biologically effective radiation, i.e. the spectrum weighted by a biological action spectrum such as that for erythema or DNA (Madronich and DeGruijl 1994).

A recent related aspect of modelling UV irradiances has been the introduction of UV forecasts, in the form of a sunburn index or a "burn time" in several countries. Indexes in different countries currently have different meanings, but there has been recent scientific agreement to standardise the indexes to the CIE erythemal action spectrum (McKinlay and Diffey 1987) and the so-called "Canadian" index. The index is then calculated in the following way: forecast meteorological fields are used to predict the column ozone for the nex day and from this the UV radiation is calculated, then weighted with the McKinlay-Diffey erythema action spectrum, integrated from 280–400 nm and divided by 25 mWm^{-2} (to give an index which is usually between 1 and 12). An empirical relationship established by Kerr et al. (1994) in Canada allows for total UV flux to be determined from total ozone for clear sky conditions. Austin et al. (1994) in the UK extended this relationship to enable direct calculation of the UV index from total ozone predictions. When compared with clear-sky measurements of UV radiation the forecast UV index had a typical error of about 9%, although the lack of aerosol in the model meant that agreement between model and measurement varied with atmospheric aerosol. Cloud is not accounted for in the model, but while some cloud can enhance UV levels for brief periods (Bais et al. 1993), clear-sky calculations should predict the maximum UV dose expected, which is an important health-related requirement. The UV index is intended to help the public adopt sensible sun exposure habits and warn them of periods of intense UV irradiance (high values of the index). An alternative expression of this information is the "burn time", i.e. the average minimum time that it would take for a fair-skinned person to begin to redden. Whichever way the information is presented, each person must learn to relate the relative values to their own skin.

1.2.2 Satellite Remote Sensing

For areal mapping and climatology purposes, satellite remote sensing has great advantages, and satellite data in the form of column ozone measurements are an important tool for solar UV studies. For years ozone measurements from the Total Ozone Mapping Spectroradiometer (TOMS) aboard the Nimbus 7 satellite were widely used, until it ceased to function in May 1993. Other satellite measurements of ozone available (WMO 1988) come from METEOTOMS (on a Soviet satellite), SBUV on Nimbus 7 and SBUV2 on NOAA11, SAGE II (onboard ERBS) and TOVS, although the signal-to-noise ratio of the latter does not encourage its use for trend detection. The SAGE II instrumentation is able to provide not only total column ozone but also vertical ozone profile information. The ozone amount is assessed from the back-scattered ultraviolet radiation from the earth measured at several wavelengths with varying ozone absorption coefficients. The relative intensities of radiation at the different wavelengths allow the ozone concentration along the radiation pathlength to be calculated. A review of satellite remote sensing of ultraviolet radiation is given by Nunez and Kalma (1995).

2 Techniques for UV Measurement

There are three basic classes of UV measurement technique, which can be used according to requirement: spectral measurements, broadband measurements and dosimetry. The first two methods generally employ instrumentation which produces and records an electric signal in response to irradiation. They are most frequently used for measuring the environmentally available radiation, defined by the radiation incident on a horizontal surface with a hemispheric, cosine-weighted field of view. Dosimetry involves a radiation-sensitive device which can be placed in a small holder (badge) and changes progressively as the accumulated radiation dose increases. The simplicity of the sensing device enables it to be used to measure the dose incident on surfaces of different orientation in situations where more cumbersome instrumentation is impractical, e.g. the human body or plant leaves. However, once the simple dosimeters have been exposed, sophisticated techniques may be required to analyse them.

2.1 Spectral Measurements

Spectroradiometry (Fig. 3) involves gathering radiation from a prescribed field of view (foreoptics), separating the radiation into its component wavelengths (monochromator), measuring the intensity at each wavelength (detector) and recording the data (logger). The control of the whole system and the data logging are tasks commonly done by computer.

For measuring solar UV radiation the standard input optics should have a hemispheric field of view with a cosine response. This is usually approximated by an integrating sphere or a diffuser of Teflon or quartz. The actual cosine responses of both types of device are often far from ideal (Fig. 4) and tend to be particularly bad at large zenith angles (low solar elevations). The cosine responses shown in Fig. 4 are measured across one plane of the diffuser (e.g. from north to south for a fixed diffuser). It is also important to know whether the cosine response is the same across all planes, or whether there is an azimuthal dependency to the cosine response which would give a daily bias to the measurements. A similar systematic bias may be incurred by incorrect levelling of the input optics, and the diffuser or integrating sphere must be mounted with great care. The imperfections in cosine responses can lead to discrepancies between measurements made with instruments having different cosine responses. Knowledge of the cosine response of a system allows the associated errors to be assessed, and in some cases a first-order correction is possible, although this requires assumptions to be made about the distribution of the diffuse radiation (usually assumed isotropic). Input optics are generally a large

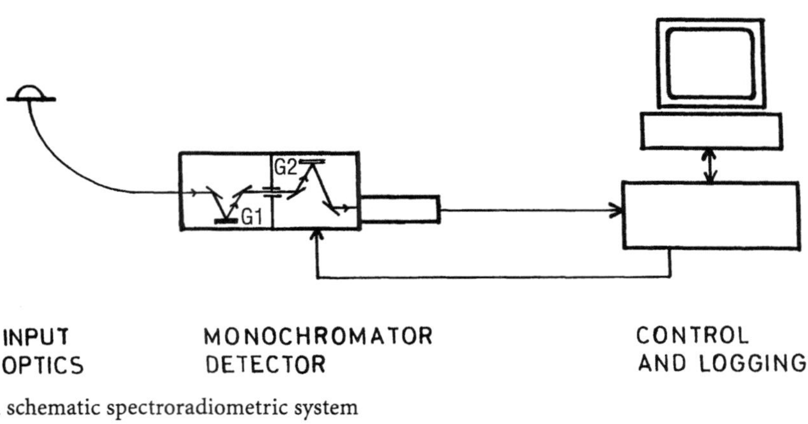

Fig. 3. A schematic spectroradiometric system

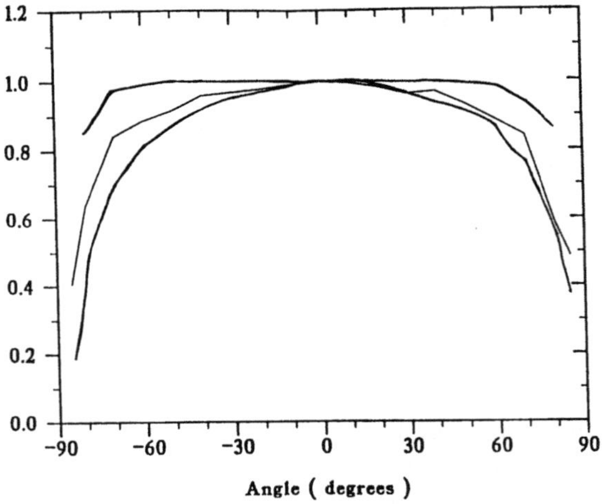

Fig. 4. Cosine responses of three spectroradiometers (data from Gardiner and Kirsch 1995). The ratios of a true cosine curve to the measured cosine responses are shown. This would be a straight line of value 1 for an ideal cosine receptor

source of error and need improvement (Seckmeyer and Bernhard 1993; Gardiner and Kirsch 1995). Foreoptics with alternative characteristics are available for specialised measurements. For example, a device with a narrow field of view (about 1°), mounted on a solar tracker, can be used to measure direct beam solar radiation.

Radiation gathered from the input optics is steered directly, or by means of a light guide, into the monochromator; the defining element of a spectroradiometer. The radiation may be separated into its component parts by either a prism or a diffraction grating, and may go through this process either once (single monochromator) or twice (double monochromator, Fig. 3). The dispersers within monochromators are predominantly diffraction gratings because their attenuation of radiation is low, they can be tailored to optimise their performance and they provide a linear angular dispersion, none of which is true of the more robust prism. It is important so select a grating which has been blazed to optimise performance in the UV region of the spectrum (typically blazed at 250 nm and 2400 lines/mm). Diffraction gratings may be either plane or concave. A plane grating will produce a curved spectral field which must be flattened by a concave mirror before it is directed onto the (straight) monochromator exit slit. A curved grating will both disperse and produce flat field radiation, but this dual tasking will result in some degree of aberration which can only be minimised over a limited wavelength range.

The shape of the solar spectrum places very stringent requirements on the performance of a monochromator (Table 1). Because the intensity increases so sharply (several orders of magnitude) across the narrow UVB part of the spectrum, the wavelength alignment and resolution must be very precise and stray-

Table 1. Requirements for a spectroradiometer to measure solar UV radiation (Gardiner and Kirsch 1995)

Feature	Specification	Determining component
Wavelength range	280–400 nm +	Grating/prism and detector
Wavelength resolution	≤1 nm	Grating/prism and slit function
Wavelength accuracy	Stable to 0.2 nm	Monochromator
Sensitivity/dynamic range	>6 orders of magnitude	Detector and overall throughput
Straylight rejection	10^{-6} $Wm^{-2}nm^{-1}$	Monochromator, slit function
Field of view	Cosine response	Input optics
Operating environment	Stable, controlled	Thermal stabilisation and weatherproofing

light must be vigorously suppressed, otherwise photons (energy) of one wavelength will be wrongly attributed to another wavelength. Wavelength alignment is achieved by precise mechanisms and careful control and should be accurate and repeatable to within 0.2 nm (Gardiner and Kirsch 1995). However, the stringency of this requirement will depend to a certain extent on the intended use of the measurements, as the error incurred by misalignment increases as wavelength decreases. At 300 nm an uncertainty of 0.1 nm in wavelength alignment can lead to an uncertainty of the order 5% in the measured irradiance at that (nominal) wavelength. Wavelength resolution is defined as the smallest incremental changes in wavelength that can be resolved and measured. It depends upon the angular dispersion and control of the grating (the wavelength step size possible) and the slit function of the instrument. The slit function is the shape of a single monochromatic line measured by the instrument. Ideally this should be a single high intensity line with no measured radiation at adjacent wavelengths. In practice it is often a triangular shape peaking at the wavelength of the radiation source. Radiation measured at adjacent wavelengths represents near-field straylight. The slit function is often specified by its full width half maximum (FWHM), i.e. its width at half the maximum intensity (Fig. 5). The smaller the FWHM the better the resolution of the instrument. A resolution of

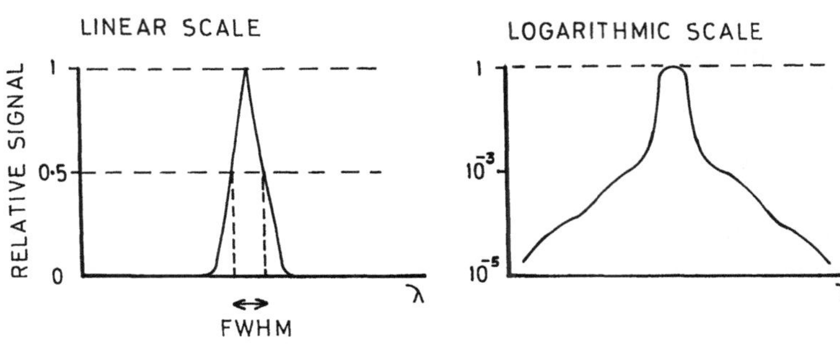

Fig. 5. A schematic representation of the slit function of a spectroradiometer

1 nm or better is desirable for solar UVB measurements, and a complete knowledge of the shape of the slit function is recommended, as broad "wings" on either side of the central peak may be important at the shortest solar wavelengths where intensity is increasing most rapidly. In addition to the slit function, straylight can also reach the detector from other wavelengths which have been misdirected within the monochromator and fall upon the exit slit when another wavelength is nominated for measurement. This is far-field straylight and can be reduced by having non-reflective surfaces in the monochromator and avoiding overfilling optical surfaces (leaving "spare" radiation as a straylight source).

The straylight properties of double monochromators are generally significantly better than those of single monochromators because of the second level of dispersion and the additional baffling involved. As straylight rejection is so critical to solar UVB measurements, a double monochromator is usually required.

The precipitous shoulder of the solar spectrum is also demanding of the detector for the spectrometer. It needs to be very sensitive, to detect the low-intensity but biologically active short wavelengths, but must also have a large dynamic range to enable the rest of the spectrum to be measured. Detectors sensitive to the UV and visible regions of the spectrum are photomultiplier tubes (PMT) and silicon photodiodes. For low-intensity spectral measurements PMTs are usually the detector of choice. If the correct PMT is selected for the UV region it will be orders of magnitude more sensitive than semiconductors in the UV/visible region and have a linear response over 6–8 orders of magnitude. However, PMTs have the disadvantage of being bulky and requiring a specialised power supply. They are also temperature-dependent and need temperature stabilisation, or characterisation and monitoring, especially if being used in variable conditions outdoors.

If a PMT (or single photodiode) is used, each wavelength of radiation must be directed in turn onto the detector. This scanning takes a finite time and can be a problem if conditions are changing. Spectral distortions due to such changes can be easily identified if a broadband instrument (not necessarily UV) is sampled at intervals throughout a scan: stability in the broadband signal indicates consistent sky conditions for the scan duration. Another possible solution is to use a diode array detector, a line of small semiconductor detectors each receiving radiation from a narrow waveband. The complete set of detector elements can be sampled by computer in milliseconds to provide a snapshot picture of the spectrum. The disadvantage of current diode array detectors is the susceptibility to straylight of the systems into which they must be built (generally single monochromators). The necessary strictures on straylight rejection, resolution and signal-to-noise ratio are difficult to meet for solar UVB measurements.

A spectroradiometer which satisfies the optical specifications for solar UV measurements must also be capable of operating in the complete range of environmental conditions to which it will be exposed. This means insensitivity to, or protection against, temperature extremes and moisture. This is often achieved by enclosure of the monochromator and detector in a waterproof, temperature-stabilised housing. The monochromator is then connected to the external input optics by a light guide (liquid or quartz fibre). Other connections are required to the control and data storage modules and to a power supply (usually mains). All this renders the solar UV spectrometer a bulky and cumbersome instrument which is transportable rather than portable.

The advantage of a spectroradiometer is the versatility of its data. The irradiance at each wavelength is available and so the potential effect of a source can be calculated for any response whose action spectrum is known. The spectral signature of absorption by given elements can be identified, and the amount of absorbing substances can be calculated by the strength of absorption at different wavelengths, e.g. total column ozone, which is best derived from direct sun measurements using a narrow field of view detector but can also be calculated from global measurements (Stamnes et al. 1991). The ability to look at narrow parts of the spectrum in detail is a great advantage, especialy for solar UVB measurements, where important information about the ozone-sensitive, biologically active short wavelengths is easily swamped by radiation at wavelengths only a few nanometers longer. The price of this information is the expense and care needed to operate a spectroradiometer, with the result that spectroradiometry has been predominantly a research tool, although the number of routine solar spectral monitoring sites is increasing (see Sect. 4, 5). Most of the spectroradiometers in use today have been purpose-built for the task or are commercial instruments modified for use outdoors and are thus unique in some mode of their operation. One exception to this is the Brewer spectrophotometer, which is predominantly used for ozone measurement but can also provide limited spectral UV data in its standard operational mode (see Sect. 5).

2.2 Broadband Measurements

Broadband radiometers measure the total irradiance contained in a specified waveband. The waveband is determined by the detector and by filters or other moderating devices placed between the radiation source and the detector, which is typically a solid-state device. Specifying the waveband of radiation over which the radiometer operates is not sufficient to understand its output; the response spectrum of the instrument must also be known. To measure straigtforward energy (the irradiance integrated across the designated waveband of operation) the response spectrum should be flat with steep cut-offs at either end of the required waveband. For UVB radiation this results in a measurement dominated by radiation in the longest 10 nm of the waveband, and thus not very susceptible to changes in UV due to ozone depletion. As the solar UVB energy is negligible at ground level (less than 0.5% of the total solar energy), there is far greater interest in the short-wavelength, biologically effective and ozone-sensitive end of the UVB part of the spectrum. To satisfy this interest many UVB sensors have a response spectrum which is tailored to imitate the erythemal action spectrum (Fig. 6) and is weighted towards the short wavelengths. The first such sensor to be widely used to monitor solar UV radiation was the Robertson-Berger meter (see Sect. 4). A variety of "erythemal" UV meters are now commercially available, all with slightly different approximations to the erythemal action spectrum, depending on the filtering technologies employed.

The output of a radiometer gives the total irradiance (weighted or not) across the waveband of the instrument response: a single number at any instant in time. This makes recording of the signal for spot samples or cumulative totals a simple task and no control of the detector is required. Thus the instrument is much smaller, cheaper and simpler than a spectroradiometer, but at the cost of loss of information and versatility in the data analysis. A broadband instrument provides no information about the spectrum of the incident radiation, and if a change in the radiation source is detected it is not possible to say whether the change is spectrally dependent or a uniform change at all wavelengths. Broadband meters with different response spectra will give different indications of a change in irradiance if that change has a spectral characteristic, making it difficult to compare meters of different design and to extrapolate measurements to apply to alternative biological effects with different action spectra. Meters of the same type can be used to identify changes in irradiance with space and time, assuming suitable calibration procedures (see Sect. 3).

The field-of-view (hemispheric, cosine response) and environmental requirements (waterproof, insensitive to temperature and humidity) for the broadband meter are the same as for the spectroradiometer. The simpler components employed in broadband radiometry mean that the instruments are often

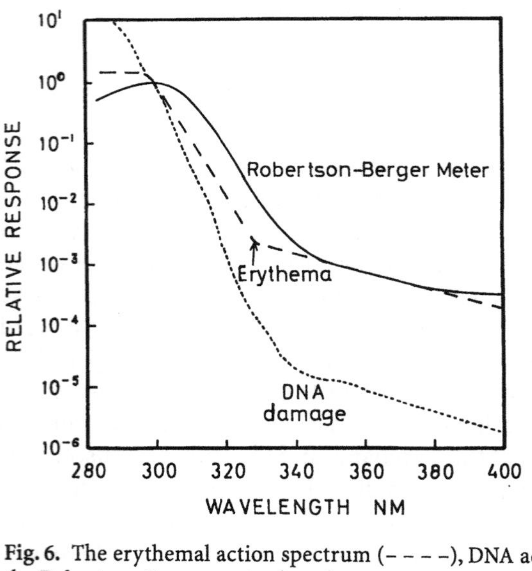

Fig. 6. The erythemal action spectrum (– – – –), DNA action spectrum (·······) and response of the Robertson-Berger meter (——)

deemed temperature-independent with no further stabilisation necessary, although some of the most recent UV radiometers include temperature control elements. Unlike most spectroradiometers, many of the broadband radiometers have been designed for long-term use outdoors and are supplied in weatherproof housings. The cosine responses are no better than those of the spectroradiometers. Despite their limitations, the ease of use of broadband radiometers has led to more extensive deployment of these instruments for environmental monitoring than the more complex spectrometers (Sects. 4, 5).

A further class of instrumentation is the multi-channel narrowband radiometer, which is somewhere between a spectrometer and a broadband instrument. It detects radiation in several narrowly defined regions of the spectrum, typically 4 or 5, at e.g. short-wavelength UVB, the UVA/UVB border region, UVA and PAR (photosynthetically active radiation). The pass band across each narrow region is typically triangular in shape, like a spectroradiometer with a wide slit function. As with the spectroradiometer, a full knowledge of the pass band functions is important as the FWHM is insufficient to assess the channel's performance in some cases. One method of labelling the pass band is to use the effective band width (EBW). This is determined by constructing a rectangular function about, and at the same height as, the central wavelength of the pass band. The area under this function should have the same area as the true pass band function, and the width of the rectangular function is then the EBW. Specifying the EBW in this way takes some account of any broad wings that the pass band may have, and could equally well be used for spectroradiometer slit functions. The multi-channel radiometers provide information on thin slices of the spectrum which, if well chosen, can have wider applications than the data from broadband radiometers, but do not give the complete information available from the spectroradiometer.

Appendix A gives a list of some of the UV instrument manufacturers.

2.3 Dosimetry

Dosimetry is the measurement of the dose of radiation received by a given surface over a specified period of time. The term is often used to imply personal dosimetry – a measurement of the exposure of a person to a particular type of radiation – but it can equally apply to a plant leaf or the wall of a building. The standard exposure of radiometers and spectroradiometers is a horizontal surface: dosimetry investigates the actual exposure of variously oriented surfaces during daily activities, or phases of growth, or different seasons. To allow its deployment in a multitude of places and situations the dosimeter should be small, require no power and be immune to temperature and humidity changes. Ideally it should require no special treatment immediately after exposure and should be capable of being stored without change for prolonged periods before evaluation.

Several dosimeters for UV radiation have been proposed and used in a variety of studies (Davis et al. 1976; Webb et al. 1988; Ronto et al. 1994). The most widely employed UV dosimeter is polysulphone film (Davis et al. 1976), a clear polymer film that can be made into small badges for epidemiological or other uses (e. g. Diffey et al. 1977; Webb et al. 1990; Herlihy et al. 1994). The film changes its absorbance at 330 nm on exposure to UV radiation and has a response spectrum similar to the erythemal action spectrum. Pre- and post-dose absorbance measurements are required and the absorbance can then be related to the erythemally effective dose of radiation received by the film badge, and the surface to which it was attached.

Another approach to dosimetry is not to measure biologically effective radiation with an approximate, fabricated action spectrum, but to use the direct biological effect to assess the radiation, giving the dosimeter a specific and precise response spectrum. Biosensors based on UV damage to DNA (Fig. 6) involve the exposure of bacteriophages (e.g. phage T7, Ronto et al. 1994) or bacteria (e.g. *E. coli,* Tyrrell 1978) to radiation, and then evaluation of the exposed population for damage. Repair-deficient strains of bacteria are used to enable storage of the samples before evaluation.

Solutions of 7-dehydrocholesterol (7DHC) have been used in effect-specific studies to determine the potential for vitamin D formation in different climates (Webb et al. 1988). The first stage of vitamin D synthesis requires the potoconversion of 7DHC to previtamin D_3 by wavelengths less than 316 nm. However, once formed, previtamin D_3 is itself photolabile and forms a quasi-equilibrium with three of its isomers which is dependent on the irradiating spectrum (MacLaughlin et al. 1982). I. Terenetskya has proposed previtamin D as an UV dosimeter which by the composition of the irradiated mixture of isomers could give spectral information about received dose as well as total exposure (pers. comm.). A discussion of UV dosimeters and their uses is given in a special edition of *Journal of Photochemistry and Photobiology B: Biology* (1995, Vol. 31, issues 1, 2).

Dosimeters suffer from the same restrictions as broadband meters in that they report only the total (weighted) result of an irradiation, often expressed in biologically effective units. The relation between two "effect units" is dependent on the spectrum of the irradiating source. The relation between the two units established under one source will not hold if the source changes: Webb (1985), calibrating polysulphone film in physical energy units under the solar spectrum, found that different calibration factors were needed in different seasons as the shape of the midday solar spectrum changed.

Madronich and deGruijl (1994) calculated the radiation amplification factors (RAF: the percent change in biological effectiveness of solar radiation for a 1% decrease in ozone) for erythema and DNA damage for the ozone data of 1979 and 1992, and found values of 0.7–1.2 for erythema, and 1.5–2.3 for DNA damage, depending on latitude and time of year. These results approximate the results that would have been obtained from dosimetric measurements with polysulphone film and the T7 phage.

The great advantage of dosimeters is that they can be used in situations where it is not possible to use radiometers. The dosimeters themselves are generally small and simple and can be used in large numbers. However, evaluation of the dosimeters often involves sophisticated techniques and can be a labour-intensive task. Nonetheless, they presently represent the only way to assess the actual exposure of mobile surfaces, and a dosimeter placed horizontally would allow for reference back to radiometric and spectroradiometric measurements.

3 Calibration Techniques

The calibration of any measurement device is vital to the credibility of the data and defines the lower limits of uncertainty associated with the measurement. The calibration procedure is the means by which information from the instrument is related to a common standard and accepted unit, with a known margin of error. The standards for many quantities are known to a high degree of accuracy and are precisely defined (e.g. m, s). This is not the case for standards of UV spectral irradiance, where even the primary standards of national standards laboratories differ from each other by several percent in the UVB region (Walker et al. 1991).

The lack of anything approaching a definitive standard is the most basic, but not the only, problem associated with efforts to standardise UV measurements. While the second can be defined to one part in 10^{13}, UV irradiance standards barely reach consensus and transferral of one part in 10^2. As no measurement can be better than the calibration on which is based, there is an inherent error of 2–3% in all UVB measurements, which is a direct transference of the uncertainty in the standard. Two instruments calibrated to different standard sources may thus make measurements that differ by up to 5% even if all other aspects of the calibration and measurement procedures are perfect. Better agreement can sometimes be achieved (Webb et al. 1994), but if different instruments calibrated on different sources from different standards laboratories are compared, expectations of agreement to better than 5% are optimistic.

The collection and comparison of data from individual monitoring stations to assess locational differences in UV irradiances must be made with an understanding of the uncertainties associated with even the most careful measurements. As most of the UV monitoring stations around the world are independent, a global, or even regional, view of the surface UV radiation must rely on variously sourced data, so assessing the level of agreement between different instruments is a vital task. The intercomparison of instruments followed by their immediate deployment at comparative sites can reduce uncertainties associated with the independent systems in the short term. This method has been used in exploring the altitude effect (Blumthaler et al. 1994b) and hemispheric differences in UV (Seckmeyer and McKenzie 1992), but in the long term the standardisation between sites will only be maintained if the calibration at each site is consistent. Internal consistency is also a prerequisite for defining the expected levels in UV (climatology) and any changes or trends away from this expectation. Section 5 will discuss the efforts at standardisation between various groups involved with solar UV measurements. Individual calibration methods are described below.

3.1 Standard Lamps

The most common method of calibration is to measure a standard lamp, i.e. a source with a well-defined output, and equate the signal from the instrument with the known source output in carefully controlled conditions. Ideally the calibration lamps should be similar in intensity and spectral shape to the unknown source(s) that the instrument will then be expected to measure. This reduces the need to assume that the instrument response is linear and straylight properties do not change with the irradiating source. The artificial source which most closely resembles the solar output is the xenon lamp combined with an appropriate cut-off filter. Unfortunately xenon lamps are not stable enough for use as standard lamps, and the choice of lamps becomes quartz tungsten halogen (QTH), deuterium, or both. The QTH lamp has a smoothly increasing output from the UV through the visible and into the infra-red, and stability is good across this broad spectral range. However, intensities in the UVB are 1–2 orders of magnitude less than with a xenon lamp or in bright sun. The deuterium lamp has a higher output than the QTH at short UVB wavelengths (the exact wavelength where the two source spectra cross over depends on the relative powers of the lamps), but its spectral output decreases continously across the UV waveband (UVC to UVA) and there is no useful output at longer wavelengths. As many solar UV spectroradiometers measure, and are therefore calibrated, at UV and visible wavelengths, the QTH lamp is the most common calibration source. If available, a deuterium source may also be used to check the calibration in the UVB region where the two lamp outputs cross over.

Standard lamps are available from various national standards laboratories, e.g. NIST (National Institute of Standards and Technology, USA), NPL (National Physical Laboratory, UK), PTB (Physikalisch-Technische Bundesanstalt, Germany), or may be "traceable to" one of these laboratories, i.e. calibrated by a certified laboratory which uses standards reverting to the specified national standard (see Appendix B). The lamps are supplied with a calibration certificate stating their output at a number of specified wavelengths, with error, and the conditions under which this output was attained. The latter details should include the distance between lamp and detector and the lamp reference point for determining this distance, the current supply to the lamp and its polarity, and the orientation of the lamp envelope with respect to the detector. Changes in any of the above operating conditions can change the lamp output and negate the calibration (Walker et al. 1987; Webb et al. 1994). A suitably darkened laboratory is also required to avoid calibration inaccuracies due to straylight from the surroundings. A typical 1000 W QTH standard lamp requires a stabilised power supply of about 8.0 A, is seasoned for use with the filament vertical, and is placed 50 cm from the detector. Unfortunately this means that the instrument may have to be turned on its side for calibration. Even if this is possible, the instrument must then of necessity be mechanically stable if the calibration is to be valid. This is frequently not the case, and a number of solutions to this problem have been sought for instruments which normally measure looking upwards. If a light guide is in use then there is no need to turn the body of the instrument and the lamp can be used in its normal orientation. Alternatively the lamp can be turned and used with the filament horizontal so that it can be mounted vertically above the instrument. For this solution to be valid the lamp should have been seasoned and calibrated in the manner in which it is to be used: turning a lamp can change its spectral output significantly. Primary standards of this orientation are not curently available from national standards laboratories, although increasing demand may soon change this situation. Secondary standards, often of lower power, are available for use with a horizontal filament.

Instruments for solar UV measurement are often painstakingly mounted in a position with a clear field of view and suitable weather protection. Moving the instrument back and forth to a laboratory is not desirable and another method of calibration, or at least a calibration check, in situ is required. For this purpose less powerful sources are often used and placed closer to the input optics to give a similar intensity to a 1000 W lamp at 50 cm. The smaller distance also allows the lamp to be mechanically coupled to the instrument, and sometimes mounted in a complete housing so that it may be used outdoors. The housing must be well designed to prevent external ventilation and temperature from affecting the temperature of the lamp and hence its output. One hundred or 200 W QTH lamps are typically used for this task in opertional situations (e.g. the NSF network, see Sects. 4,5). NPL will supply primary standard lamps at these lower wattages, although operators frequently calibrate their own low-power external-use lamps against their primary standards. Due to the nature of their use and the common lamp-to-lamp calibration transfer process, the errors associated with in situ calibrations are expected to be greater than for the more traditional dark-room methods.

In certain circumstances the sun itself can be used as a calibration source using the Langley method (Kremser et al. 1984). This relies upon having atmospheric conditions which do not change during (at least) a full half day, so it is only practical where the climate guarantees frequent clear skies and there are no regional sources of pollution (i.e. remote sites). The method is most suited to direct-beam calibrations and has been successfully compared with laboratory calibrations for direct spectral measurements at the clean-air site in Izana, Tenerife (M. Blumthaler, pers. comm.).

The sun is more frequently used in spectroradiometry for wavelength alignment. The positions of characteristic features in the solar spectrum are reasonably well known and the Fraunhofer lines can be used to determine whether a spectral instrument is correctly aligned on a specified wavelength. The uncertainties in this method stem from the choice of the reference solar spectrum. The wavelengths at which the Fraunhofer minima may be found in the different reference solar spectra vary by 0.1 nm or more at the ground, and extra-terrestrial intensities also depend upon the reference chosen (Nicolet 1989). Alternatively a mercury lamp is often used for wavelength alignment in the UV and requires no special power supply or facilities.

Standard lamps are the major method of calibration for solar UV spectroradiometers. However, they are not a good choice for the calibration of broadband instruments because their spectral output is so different to that of the sun and must be convolved with the instrument response spectrum to provide a calibration. Broadband instruments are therefore frequently calibrated against another instrument, although the calibration must eventually revert back to a standard source.

3.2 Standard Instruments

An instrument with a well- (better-)known calibration can be used as a reference against which to calibrate a second instrument. It must be possible to allow both instruments to view a source in an identical manner and at the same time (if the source is not extremely stable). The output of the source does not have to be well established or even constant during the calibration if the instruments are well synchronised. The latter is harder to achieve with a spectroradiometer than with a broadband instrument, but it is possible (Blumthaler et al. 1994b). The sun is therefore a suitable source for this method of calibration, and the interplay of instrumental features (e.g. weighting function) and the solar spectrum is automatically included in the calibration. However, it should be recognised that the relation between instruments with different weighting functions will change with a change in the source spectrum.

Instruments which have been mounted for perpetual monitoring may be left in situ and the standard instrument placed close by while simultaneous measurements are made. The irradiance values from the standard instrument are then compared with those of the instrument under test as a check, or used to calculate new calibration factors from the raw signal of the test device.

The standard instrument must be well characterised and calibrated and must have proven stability over the intervals between its own calibrations. A spectroradiometer may be used to calibrate a broadband radiometer if the reponse spectrum of the broadband meter (or the response that it is to represent) is well known so that the spectral data can be correctly weighted and integrated. If the actual and intended response spectra are greatly different then the calibration will only be valid in conditions of incident radiation similar to those during the calibration (i.e. same latitude and season if the sun is the source). This is equally true if one broadband meter is used to calibrate a second with a different response spectrum.

The standard instrument calibration is most useful when a set of nominally identical instruments are deployed in a network for continuous monitoring. The monitoring instruments do not have to be disturbed and a standard instrument can be circulated amongst the sites providing regular calibration checks so that all the operational devices are referenced to the same standard. The standard instrument is then the only one which requires frequent checking at a higher level, either against better defined instruments or standard lamps. No special calibration facilities are needed at any of the other sites. Many of the broadband networks (Sect. 5) e.g. Australia, UK and America, use the standard instrument calibration method. Broadband meters are small and easily transportable, stable and mechanically rugged, and straightforward to install and operate at different sites, and so lend themselves to a peripatetic lifestyle.

A portable standard spectroradiometer is a more complex proposition. There are no large families of identical instruments and no one spectrometer that stands out as an obvious standard. All the spectroradiometers are transportable rather than portable and require careful handling and trained operators, which would greatly increase the cost of having them continously on the move. There are standard Brewer spectrophotometers used to check all the functions of Brewers around the world, but they are not used as standards with other spectroradiometric systems and have a limited wavelength range. Where a Brewer is used for serious UVB measurements, standard lamps are also used for calibration of the absolute irradiance measurements.

Spectroradiometers (independently calibrated) have been compared at the same site to establish their relative performances (see Sect. 5) and a portable suite of standard lamps in a housing for outdoor use has been proposed as a means of referencing measurements made at disparate sites to a single standard (see Sect. 5). In the meantime spectral data are referred to the standards of many different national laboratories through independent calibrations at the home site.

3.3 Frequency of Calibration

Whichever method of calibration is used, the task must be performed with sufficient frequency to maintain an acceptable level of uncertainty in the data. If a calibration changes from one determination to another, then all the data gathered in the interim period have an uncertainty associated with the difference between the calibrations, as it is not possible to say when, or how gradually, the change occurred. The calibration frequency will depend upon the stability of the instrument as a whole and the way in which it is used. In general, a spectroradiometer with its constantly moving parts and additional requirement for wavelength alignment requires more frequent calibration checks than the simpler, fixed-parts broadband meters. Transporting an instrument is more likely to incur calibration changes than leaving it in one position. An understanding of the possible causes of changes in calibration, and experience, will indicate the calibration frequency necessary for each system. All systems will change with time, if for no nother reason than a weathering of the outermost layer of the input optics.

The wavelength alignment of many spectroradiometers is frequently checked, often before each scan, as this can be done using the Fraunhofer lines of the sun (e.g. Gardiner and Kirsch 1994). Absolute irradiance calibration may be checked using an internal lamp, as with the NSF network, and a more rigorous calibration performed if a change is indicated, or as part of a pre-determined routine. In other systems an external lamp is always used as a standard, and there may be a hierarchy of lamps that are used with decreasing frequency as their specifications improve, e.g. the sites at Reading and Garmisch-Partenkirchen (section 5).

Broadband instruments sometimes go for long periods without recalibration, especially if groups of instruments rely on a travelling standard to perform the service. While this may be warranted by the stability of the systems, it has lead to criticisms of some early time series of measurements. The findings of the US network in the 1970s and 1980s (Sect. 4) were questioned because of infrequent calibration over the 11 years of the study, and the suspicion of instrumental drift which would account for the observed decrease in UVB radiation. However, it has recently been found that the type of lamp to which the calibrations of the instruments was referred was changed during the period in question, causing a shift in the calibration of the network (I. Deluisi pers. comm.). It was this shift rather than a gradual drift which caused so much speculation about the reported measurements. A greater awareness of the need for extremely stable instrumentation, and frequent, independent checks of performance if trend analysis is to be attempted, has led to improvements in instrument design and more attention to regular calibrations. Whatever the instrument, a calibration routine should be determined to suit the system stability, facilities available, measurement timetable, and the error tolerance in the type of data analysis expected.

The calibration procedure, the instrument characteristics, and the conditions (site) under which measurements are made will all contribute to the uncertainty associated with a measurement. It is vitally important that the uncertainty in a measurement is assessed so that false assumptions of accuracy are not made in using the data, and data may be matched to applications with appropriate uncertainty tolerances (see Sect. 6). The construction of an uncertainty budget, and the definition of all the terms involved, is a complex subject which differs for each class of instrument. At the time of writing, guidelines for error analysis are being prepared by the WMO Scientific Steering Committee for UV Measurements (see Sect. 5). Similar guidelines for the requirements of UV measurement sites intended for different porposes are also in preparation. The reader is directed to these publications which will cover these important topics in more depth than is possible here.

4 The History of Measurements

The widespread measurement of solar UV radiation for its own sake is a relatively modern concept which first aroused public interest in the 1970s and 1980s with concern for the ozone layer, first in connection with proposed fleets of supersonic aircraft (CIAP 1974; COMESA 1975), and then, with more vigour, when the Antarctic ozone hole was publicised in 1985 (Farman et al.) Before that the UV, and particularly the UVB, part of the spectrum at the surface had been largely ignored as containing little energy, although there had been discrete studies of the relation between ozone and UV, and the determinants of UV radiation at the surface. Details of the earliest UV measurements, many of which lacked units and are difficult to interpret, can be found in Webb 1996.

Solar UV radiation has long been used to assess the amount of ozone in a vertical column of the atmosphere. Dobson (1931) first proposed his spectrophotometer in 1931, using pairs of UV wavelengths (one strongly and one weakly absorbed by ozone) to assess the amount of atmospheric ozone. The instrument could be used with direct sunlight, or with zenith skylight, and became the standard instrument for the ground-based ozone monitoring network. The longest record of Dobson ozone measurements can be found at Arosa (Switzerland) and dates back to 1926, but few of the early records extend unbroken until the present day. With the International Geophysical Year in 1957 the number of monitoring stations was increased and run as a network by WMO (The Global Ozone Observing System, GO_3OS). The number of ozone monitoring stations has continued to increase, but truly global ozone data were not available until the advent of satellite measurements in 1979.

While the Dobson spectrophotometer (Dobson 1931) uses UV radiation, it does not rely on absolute measurements, but on the relative intensities of pairs of wavelengths. Thus no UV data is obtained from the Dobson network. In recent years the Brewer spectrophotometer (Brewer 1973; Appendix A) has become increasingly widespread as an alternative to the Dobson instrument for ground-based monitoring of ozone (and other atmospheric gases). The Brewer works on the same basic principle as the Dobson but includes the capability of making absolute spectral UV measurements in the range 290–325 nm. Not all ozone stations make use of this facility, but there is an expanding potential for UVB measurements whenever a Brewer spectrophotometer is employed for ozone monitoring.

As the ozone monitoring network was becoming established, Bener was beginning several years of dedicated spectral UV measurements at Davos in Switzerland (Bener 1960, 1963, 1964). His purpose-built spectroradiometer, at a high-altitude site, produced detailed information on the UVB and UVA radiation reaching the surface, and provided the knowledge and understanding of factors affecting UV irradiances (Bener 1969) which have been the basis of UV-ozone work ever since.

The next attempt to measure solar UV radiation represented the increasing realisation that UV radiation has detrimental biological effects. The increasing incidences of skin cancer prompted a desire to know more about the distribution of the main causative effect, UVB radiation. The Robertson-Berger meter (Berger 1976) was designed to measure broadband UV radiation weighted with an approximation to the erythemal action spectrum. The meter reported "sunburn units" and was deployed in networks in Australia (Barton and Paltridge 1979) and the USA which date back to the 1970s. In 1988 a report from the US network (Scotto et al. 1988) concluded that the amount of incident erythemally effective UV radiation in the USA had decreased between 1974 and 1985, in contradiction to the ozone decrease that had been observed over the same period. The validity of the data was questioned for a number of reasons; the siting of the meters (often in areas where atmospheric pollution had increased over the years of the study), and the calibration and stability of the instruments. A number of studies of Robertson-Berger meters and data since then have not enirely resolved the issues raised (DeLuisi et al. 1992; Kennedy and Sharp 1992; Frederick and Weatherhead 1992; Justus and Murphy 1994). Another study using the same instrument at a high mountain site in Switzerland (Blumthaler and Ambach 1990) reported an increase in UVB radiation between 1981 and 1989 of 0.7% per decade, more in accordance with the ozone decreases being observed. Studies from Russia (Garadazha and Nezval 1987) reported a 12% change in surface UVB in the 15 years from 1968 to 1983. In the last few years the Robertson-Berger meter has been updated to give a better approximation to the erythemal action spectrum, and now also includes temperature stabilisation. This new unit is known as the Solar UV Biometer and is just one of several instruments now available for measuring solar UVB radiation.

The Australian network was not maintained thoroughly in the 1980s, but more recently another network has been established at major centres of population using a different model of erythemal sensor (International Light, Appendix A) and some of the new version Solar UV Biometers. Similarly, New Zealand has a network of broadband sensors, which recently reported an upward trend in UV radiation from a station at Invercargill (Zheng and Basher 1993). Robertson-Berger/Solar Biometers are used in networks in Germany, the UK, the Scandinavian countries and at many other independent locations.

A Canadian company (Vital Technologies Corp., Appendix A) manufactures another broadband erythemal meter which is being deployed in schools in Canada as part of an educational network. The Vital instrument is also deployed in a network being developed by Bausch and Lomb in the United States to provide public information. This network also uses a multi-channel narrowband instrument from Biospherical Instruments Inc.

Broadband UV monitoring sites and UV-related investigations continue to increase, albeit as a series of discrete efforts rather than a coordinated task. Unfortunately there is little information about pre-ozone-hole climatology. The longest time series of UV irradiances at the ground comes from a multi-filter instrument deployed in Maryland, USA, from 1975 to 1990 (Correll et al. 1992). The data show a large increase in UVB, especially at shorter wavelengths, over the period 1980–1987. After 1987 the irradiances decreased, which may be due to either changes in the instrument made at this time, or changes in clouds, aerosols, etc.

Spectral details of solar radiation are even harder to come by than broadband data. For 20 years after Bener's work there was little in the way of spectral UV investigations, and it is only in the last 5 years that routine spectral monitoring has been established at a small number of sites worldwide, with investigations but not continuous measurements at a few other locations. There are reports in the literature of measurements made for a limited period of time in the 1970s and 1980s (e.g. Tarrant and Brock 1975; Wester 1984; Webb 1985; Hisdal 1986; Webb and Steven 1987; Webb and Campbell 1987), but there are no long-term pre-ozone-depletion data with which to compare the measurements of today.

The National Science Foundation (NSF) of America established a network of spectroradiometers in Antarctica in 1988. Three identical instruments were deployed at bases at the South Pole, Palmer and McMurdo stations. A fourth base in Ushuaia, southern Argentina, was added in 1989, and two northern hemisphere sites at Barrow, Alaska, and San Diego, California, became operational in 1990 and 1992 respectively. The instruments are built, and the network run, for NSF by Biospherical Instruments Inc. (Apendix A). Measurements are made from 280 to 600 nm at wavelength-dependent resolution every hour. The long-wavelength measurements serve to indicate the general state of the atmosphere (e.g. cloudiness) and its effect on radiative transfer. If the short-wavelength UVB is anomalous but UVA and visible radiation are normal for the time of year, changes in atmospheric ozone may be implied. Such observations have been made at the Antarctic stations during the spring ozone hole periods when unusually low ozone over the stations is mirrored by correspondingly high UVB (Lubin and Frederick 1989; Booth et al. 1994). High UVB has also been observed at Ushuaia when the ozone hole has on occassion extended as far north as southern Argentina (Frederick et al. 1994).

The Japanese Meteorological Office runs a series of four Brewer spectrophotometers, spread latitudinally through the islands of Japan, which have been collecting spectral UV data as part of their operational programme since 1990 (Ito et al. 1993).

Other spectral UV monitoring sites have, to date, been independently established and use a variety of different instrument designs, although there are informal associations between some of the sites (Sect. 5). In the southern hemisphere there are spectral measurement facilities at Lauder, New Zealand (McKenzie et al. 1992), at Melbourne, Australia, at Cape Grim, Tasmania, and at Durban, South Africa, which have been operational during the 1990s. Data from Lauder have been compared with data from Neuherberg, Germany, at a similar latitude, to explore differences in UV radiation in the two hemispheres. In addition to the differences in column ozone between the sites (less at Lauder), other factors also made some contribution to the higher UVB irradiances observed at Lauder: the cleaner southern atmosphere and changing earth-sun distance (smaller during the southern summer than during the northern summer) (Seckmeyer and McKenzie 1992).

The changes in ozone in the northern hemisphere have been far less dramatic than those over Antarctica. The natural variability in not only ozone but also cloudiness and pollution means that the current records of spectral measurements are not long enough to provide unequivocal evidence of generally increasing UV irradiances at the surface. However, in 1992 and 1993, when there was unexpectedly low ozone for prolonged periods over parts of the northern hemisphere (due to the effects of Mount Pinatubo and anomalous weather conditions (Stowe et al. 1992; SORG 1993)), elevated levels of UV radiation were reported from Germany (Seckmeyer et al. 1994) and Toronto (Kerr and McElroy 1993). The Toronto measurements – part of a series dating back to 1989 – were analysed to show that the spectral characteristics of the high irradiances had the distinctive signature associating them with anomalously low ozone. Here and elsewhere, measurement records are not long enough to warrant any discussion of trends in UV irradiances. However, spectral measurements are also being used to explore some of the other factors that affect UV radiative transfer and incident irradiances, e.g. altitude (Blumthaler et al. 1994b), albedo (Feister and Grewe 1995), and cloud (Blais et al. 1993).

5 Networking

Figure 7 shows the current status of UV monitoring worldwide. The distribution of instruments is uneven, and over much of the land surface it is extremely sparse or non-existent. The North America continent and western Europe account for approximately 65% of all the instrumentation. Elsewhere, Australia, New Zealand and Antarctica are covered in the regions of major population centres and over a range of latitudes. Asia, Africa and South America are very poorly represented, especially as, although the Brewer spectrophotometers shown on the map have the potential for UVB measurements, they are not necessarily in operational service in this mode. Where data are available they come in various formats. There are broadband (usually erythemally effective) data represented by a small number of established instrument types, which within each group are nominally the same and so directly comparable. The spectral data are far more diverse since both the instrumentation and operating protocols are more individual and there is no established format for data presentation.

The scarcity of data as a whole makes it desirable to be able to use all the available information as a single database. The logistics of acquiring output from all the monitoring sites and data reduction to a common format is a task waiting to be tackled. On a more fundamental basis it must be proved that there is a relationship between data from various sources: that if instruments are placed side by side in the sun they can produce the same irradiance values in the same format (units), and that moving apart they can maintain this agreement. Such an assurance is expected of the few formal national networks of UV instruments, and is sought informally by groups of independent operators on an inter-

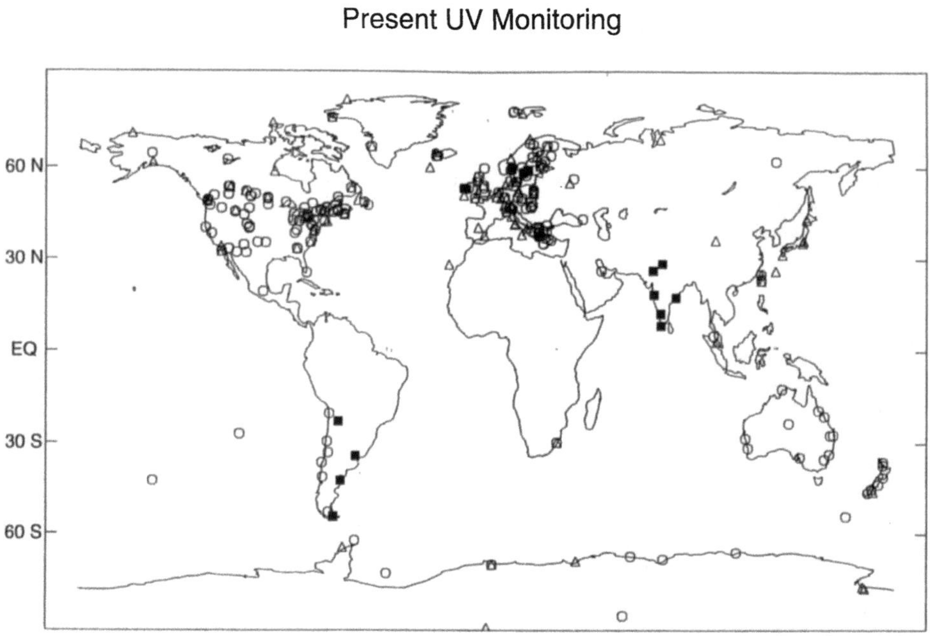

Fig. 7. The global distribution of solar UV monitoring sites showing broadband (o), spectral (including Brewer spectrophotometers) (Δ), and multi-filter (■) instruments. Only instruments taking routine measurements are represented. (Data supplied by Dr. E. C. Weatherhead, November 1995)

nationals basis. Where commercial instruments are used at individual sites there may be services available from the manufacturers which will serve to maintain consistancy between all their products, e.g. the standard Brewer spectrophotometers used to check the calibration of Brewers deployed worldwide, and standard Robertson-Berger meters (or solar UV biometers) used in the same way. These services do not guarantee any standards of operational procedure and are not obligatory, but they can add to the confidence and coherence of data gathered from a family of identical instruments.

5.1 Formal Networks

The simplest and cheapest instruments to deploy in large numbers are the broadband meters. National networks of broadband UV meters in existence today are shown in Table 2 together with the number of meters involved and the year that operations commenced. The availability of the data, and their format, varies from network to network, as does additional information, e.g. National Radiological Protection Board (NRPB) in the UK also measure UVA and visible radiation with broadband meters adjacent to their UVB detectors. The first

Table 2. Broadband UVB networks currently in operation[a]

Country	No. of sites	Date began
Australia	14	1989
Canada		
AES (Atmospheric Environment Service), Vital	23	1993
Weather network SL501	24	1992
Trent University, Vital	5	1990
Chile	8	1992
Czechoslovakia	3	1996
Finland	11	1994
Greece		
University of Thessaloniki	9	1991
University of Athens	6	1992
Hungary	4	1994
Mexico	3	1991
New Zealand	9	1989
Poland	6	1993
South Africa	3	1993
Sweden	12	1991
UK	6	1988
USA		
USDA (United States Department of Agriculture)	11	1993
Bausch and Lomb	8	1992

[a] Data compiled an supplied by Dr. E. C. Weatherhead. Further details of the sites and contact persons are available on request. Additions or corrections to the table are also welcome. Correspondence should be sent directly to Dr. Weatherhead, fax: (+)1 303 497 6546; e-mail: betsy@srrb.noaa. gov

operational network of spectral instruments was that run by Biosperical Instruments for NSF in Antartica (3), Southern Argentina (1), Alaska (1) and California (1). The spectroradiometers were designed to investigate UV radiation levels and operate in the inhospitable conditions of the Antarctic continent. Data are available as weekly updates during the austral spring ozone hole period and show the short wavelength UVB, erythemal- and DNA-weighted doses, and visible radiation for the current and previous years. The visible radiation serves to indicate when pertubations are due to more general atmospheric conditions of cloudiness (all irradiances reduced) or to ozone changes (a UV extreme with normal visible radiation). At the end of a season the year's data are collected together and are available on CD-ROM. More recently other spectral networks (including the implementation of the UV option for Brewer spectrophotometers) have begun to gather data, although some are still in design and trial stage. Table 3 shows the current spectral UV monitoring networks.

Between the two classifications of broadband and spectral instruments are the narrowband, multi-channel instruments which measure over a number of discrete narrow waveband channels, thus providing some spectral information, but not the high-resolution data of the spectroradiometers. A number of countries have installed networks of this type of instrument (see Table 4).

Table 3. Spectral UVB networks currently in operation[a]

Country	No. of sites	Date began
Antarcitca/USA	6	1988
Belgium	3	1993
Canada	10	1992
China	2	1993
Denmark	3	1992
Germany		
DWD (Deutscher Wetterdienst)	5	1993
IFU (Fraunhofer Institut für Atmosphärische Umweltforschung)	3	1992
BfS (Bundesamt für Strahlenschutz)	4	1993
Greece	3	1991
Iceland	4	1993
Italy	7	1992
Japan	6	1991
Netherlands	3	1993
New Zealand	3	1991
Norway	5	1991
Poland	2	1992
South Africa	4	1994
Spain	3	1991
Taiwan	2	1991
USEPA (United States Environmental Projection Agency)	6	1994

[a] Data compiled and supplied by Dr. E. C. Weatherhead. Further details of the sites and contact persons are available on request. Additions or corrections to the table are also welcome. Correspondence should be sent directly to Dr. Weatherhead (see Table 2 footnote).

Table 4. Multi-band UVB networks currently in operation[a]

Country	No. of sites	Date began
Argentina	4	1994
Chile	3	1994
India	4	1980
Norway	7	1995
Sweden	3	1987
USA	14	1994

[a] Data compiled and supplied by Dr. E. C. Weatherhead. Further details of the sites and contact persons are available on request. Additions or corrections to the table are also welcome. Correspondence should be sent directly to Dr. Weatherhead (see Table 2 footnote).

5.2 Informal Networks

Informal networks may evolve in two different ways, almost by default or by the concerted effort of several individuals to work together within the constraints of their separate organisations. The default method involves buying a recognised instrument and using it in the recommended manner, taking advantage of any calibration facilities available. Such careful operation should allow comparison of the data from all sites with the same brand of instruments working in similar manner. While there is no obligation to disseminate any data, a hardware network of, for example, Robertson-Berger type meters exists. In the future the information highway may enable all these sites to submit data to and extract from a central database (see below), thus linking together currently isolated measurement sites.

Where instruments are modified or purpose-built for a task, the potential to extend the applications of the data beyond the immediate surroundings requires more active participation. Recognition that the combination of data from individual sites is a desirable goal has lead to several international cooperations designed to facilitate the pooling of data (within the European Union, the Nordic countries, America, the Antipodes, and intercontinental). These efforts have concentrated predominantly on spectral measurements, where the diversity of instruments is greatest, as are the potential causes of discrepancy and the versatility of data collection and presentation.

The verification of the coherence of variously sourced data is a two-stage process: (1) proving that the instruments can obtain the same results in front of the same source and (2) finding a means of maintaining or reaffirming that agreement when the instruments are separated in time and space. The best way to achieve stage 1 is to have an intercomparison with several instruments side by side measuring the sun. A number of international intercomparisons of UV instrumentation have been held (Sect. 5.3), which served in the first instance to illustrate the discrepancies between many of the systems. Where a small number of instruments have proved compatible they have been used for specific investigations (Seckmeyer and McKenzie 1992; Blumthaler et al. 1994b). There are currently too few matching spectral instruments that have been involved in routine measurements for a significant period of time to provide even an informal databank or UV climatology.

Nevertheless, with an ever increasing number of instruments with improved performance, the means to achieve stage 2 is also being explored. The two alternative approaches are to provide either a travelling standard instrument or a travelling standard lamp (see Sect. 3). A travelling standard lamp has been designed and is being tested within the European Union programme (Gardiner and Kirsch 1994). Based on a set of five standard lamps for internal consistency, it is built into a housing which enables it to be mounted over any spectroradiometer in a unique position once an instrument-specific mounting plate has been built. Radiation from a vertical filament lamp is reflected so that a vertical beam exits the lamp housing. The calibration unit can therefore be used in situ and an instrument does not have to be disturbed from its monitoring position to be calibrated. As the distance from lamp to input optics is different for each instrument the unit is intended as a relative calibration check. On each circuit of the unit an operator will ascertain that his own measurement of the unit has not changed and that his instrument is therefore stable with respect to a communal standard. As absolute calibrations must be performed by each individual site the reference unit is a test of a combination of the instrument stability, the stability of the local standard lamps, and the local facilities and operational procedures used.

In the United States NIST are working on ideas for a similar transportable lamp standard. A communal calibration and instrument test facility is also being built in Boulder, Colorado, under the umbrella of NOAA/NIST and with funding from USDA (United States Department of Agriculture), USEPA (United States Environment Protection Agency), and NSF (National Oceanic and Atmospheric Administration). It is hoped that these facilities will soon be available to the whole UV measurement community.

5.3 Intercomparisons

A number of intercomparisons of instruments for solar UV measurement have been held in recent years as the number of organisations making measurements, and the requests for data, have increased. The intercomparisons have concentrated largely on spectroradiometers and absolute measurements, although the Nordic intercomparison included total ozone instruments, filter and broadband instruments (Koskela 1994). Major international intercomparisons which have been held in the past few years include three with European Union support: Thessaloniki, Greece, 1991 (Gardiner and Kirsch 1992; Gardiner et al. 1993) and 1992 (Gardiner and Kirsch 1994), and Garmisch-Partenkirchen, Germany, 1993 (Gardiner and Kirsch 1995): New Zealand, Australia and Germany meeting in New Zealand 1992 (McKenzie et al. 1993) and with an NSF instrument in Garmisch-Partenkirchen 1994 (Seckmeyer et al. 1995); and a gathering of Nordic instruments in Izana 1993 (Koskela 1994). A campaign in Boulder, North America, in 1994 brought together the NSF instrument networks (B. Weatherhead, pers. comm.). Also in September 1994 a number of instruments from the previous European Union campaigns met in Brussels for a laboratory intercomparison to explore calibration procedures and instrument characterisation in more depth. During 1995 major field intercomparisons of spectral instruments were again held in Europe (Ispra, Italy) and the USA (Boulder, Colorado), and an intercomparison of broadband sensors was held in Finland (Leszczynski et al. 1997). The results and conclusions of these latest intercomparisons are currrently under analysis and are not yet available in the literature.

Of the intercomparisons for which reports are available the largest was the Nordic campaign, with 14 spectroradiometers, 5 broadband and 3 multi-channel instruments. Seven of the spectroradiometers were single-monochromator Brewers and an eighth was a Mark III double-monochromator Brewer. All but one of the other spectral instruments were double monochromators, 4 from Optronic Laboratories and one from Macam Photometrics (see Appendix A). The University of Tromsø operated its own design of spectroradiometer. The main results of the campaign are summarised below (Koskela 1994):

1. The agreement between the standard ozone instruments was generally excellent.
2. The secondary UV standards were able to transfer a calibration within ±3% in most cases.
3. Most spectroradiometer UV calibrations agreed in laboratory conditions to within ±4–7%.
4. Deviations in measuring the electric current caused an uncertainty of ±1% in lamp irradiance.
5. The outdoor spectral measurements agreed at least within ±15% at wavelengths longer than 310 nm but deviated more at shorter wavelengths.
6. The integrated outdoor dose rates of most of the spectroradiometers generally agreed within ±10% and those of the other instruments within ±15%.
7. The consistency of the UV instruments could be increased by use of a travelling lamp standard.

These results are broadly similar to those of the European Union campaigns. The last results of these to be published were from the 1993 intercomparison which involved 12 spectroradiometers, all of them different. A single (Mark II) and a double (Mark III) Brewer were the only off-the-shelf instruments. Two models from each of Optronic Laboratories and Bentham Instruments were represented, although they had all been modified to a greater or lesser degree by their operators. The other six instruments had been designed and built by their operators. The 1993 campaign was evidence of the tremendous improvement in some of the instruments and in the intercomparison procedures and protocols since the first exploratory meeting in 1991. One of the main innovations was completely synchronised scanning (i.e. matching the sampling times at every wavelength step). This enabled comparisons to continue in chaning weather conditions and thus increased the amount of data suitable for analysis. Synchronised scanning was subsequently used in the Nordic intercomparison, the campaigns of 1994, and an investigation of altitude on UV irradiances (Blumthaler et al. 1994b). Another advance was the use of computational techniques to correct for any slight wavelength shifts from the measured spectra and to remove the effect of the interaction of different slit functions from the ratios of scans by the different instruments.

Some of the 1993 instruments were participating in an intercomparison for the first time, and especially for the custom-built instruments this provided a rigorous test of their performance. The standard to which they were working was set by more experienced operators and their instruments who had had time to refine their equipment and operational procedures. This campaign (Gardiner and Kirsch 1995) showed that with experience it is possible to:

1. Measure the solar spectrum to ±10% at wavelengths greater than 300 nm (better agreement was achieved between selected pairs of instruments, and the UVA agreement of the group was generally better than in the UVB). It should be noted that these figures refer to slit-function corrected values, or the underlying ratio of measurements when the high-resolution variations caused by the different slit functions is disregarded. It also represents the best that was achieved by a subset of the instruments, and not the norm of all measurements.
2. Achieve results within 3% of the calibration certificate of an idependent lamp in the UV region. The recommended goal is agreement to ±3% at wavelengths greater than 300 nm, and ±5% at shorter wavelengths. It was noted that successive scans could differ by 2–3% when two well-maintained lamps were measured by one instrument, and further investigation is needed to ascertain the reason for this.
3. Perform a first-order correction for the slit functions of instruments if the slit function is well known, improving the agreement between instruments with different slit functions.
4. Perform a first-order correction for cosine response, if the response of the instrument is known. In general this also improved agreement.

The agreement between instruments with the sun high in the sky was not always maintained at lower solar elevations, i.e. there was a diurnal variation in the ratio of two sets of absolute measurements. It was not possible to completely determine the reasons for this as cosine response, levelling, position on the platform, or drift of calibration (e.g. with temperature) may all have contributed to the observed changes. An improvement in cosine responses and installation of the input optics should improve diurnal consistency.

The importance of excluding straylight from the calibration room, and high-precision current monitoring were emphasised, as in the Nordic intercomparison.

The smaller intercomparison in New Zealand in 1992 (McKenzie et al. 1992) showed good agreement between spectroradiometers from New Zealand and Germany once both had been corrected for their cosine responses. Data from the instruments at their home sites were used to determine hemispheric differences in UV radiation (Seckmeyer and McKenzie 1992). These two operators met again in Garmisch-Partenkirchen in 1994, with an Australian and an NSF instrument. The latter was an intercontinental campaign and expanded upon earlier inter-regional cooperation. Earlier, a spectroradiometer from DSIR (Department of Scientific and Industrial Research), New Zealand, took part in the 1992 European intercomparison, and a British instrument has travelled to San Diego for comparison with the fixed NSF instrument here. In this way links between spectroradiometers around the world are gradually improving, but there is still room for improvement in the agreement between different instruments.

5.4 Plans for the Future

The measurement of solar UV radiation around the world has been a haphazard affair and has depended to a large extent on the personal interest of a small number of individuals and, more recently, a few national or commercial bodies. There is no central facility that can supply UV data, and each operator chooses to encourage or restrict dissemination of their data. This is despite the fact that ozone depletion and exposure to UV radiation are international issues.

In 1992 the Atmospheric Environment Service of Canada offered to administer a UV database along the same lines as the world ozone database. Operators were free to submit their data to Toronto, from whence it would be disseminated to interested parties. The data have generally not been forthcoming and there is still no single body to approach in a search for UV data.

The World Meteorological Organisation (WMO) has recognised this problem, and in summer 1994 undertook to oversee a UV Scientific Steering Committee (SSC) to coordinate UV measurements on an international basis. The SSC aims to provide guidelines for calibration procedures, quality assurance and quality control. Information about such procedures at each site, site details, etc., should be provided with the UV data. The WMO will implement the communications infrastructure and data-handling routines needed to facilitate easy submission and retrieval of data, and in this way it is hoped that more data will become readily available to the user communities.

Practical help is promised from an intercomparison and calibration site being built at Boulder, Colorado. This is intended as an international facility where scientists can use equipment not always available at home to test their instruments in a recognised manner and compare with other UV measuring devices. The travelling lamp standards under design in Europe and the USA have already been mentioned in Sect. 5.2.

6 Using UV Data

Anyone wishing to use UV data must first decide on what information is required, then ascertain whether it is available from elsewhere or whether specific local measurements are necessary, and then decide whether or not the right measurements are within budgetary restrictions.

The major choices of data type relate to: spectral or broadband (with a specified response spectrum); general (climatological and regional) or specific (here and now); totals or averages over a period, extremes (maxima and minima), or exact values at set times; long-term or short-term requirements.

It is apparent from Fig. 7 that measured data are not available for large-scale (global) climatology, and these requirements can only be satisfied by models which have been verified against measurements at a few sites in specific conditions (usually clear skies, see Sect. 1.2.1). Past and future changes in this climatology can be estimated by changing input parameters to the model, e.g. column ozone. Models are available to provide both spectral and, directly or indirectly, broadband data.

If more exacting information is required about UV radiation in real conditions then measurements become necessary. A nearby monitoring site or a regional network will suffice it typical daily and seasonal extremes or totals are needed for the general area and climate conditions. Many sites that do measure UV will make their data available on request, but at present the only way to access the data is by direct contact with the organisation concerned. When using data from other people, it is wise to ascertain the expected margin of error in the data, the calibration methods used and the frequency of calibration, and details of the site – the horizon, proximity to pollution sources, local albedo, etc.

To be more specific, measurements must be made on site and the desire for information must be balanced with the available budget. Broadband measurements are cheaper than spectral, both in initial start-up costs and in maintenance and man-hours during operation, by a factor of 10–100. Spectral instruments need more power, more space (especially when temperature-stablised and weatherproofed), but provide more data. If erythemally effective radiation extremes or totals are the only information of interest, then one of the commercially available broadband meters (see Appendix A) with a suitably programmed logger would suffice. Where spectral data are necessary, a full spectroradiometer may be purchased (Appendix A), or with time an instrument can be constructed from component parts. Even the full commercial systems generally require some modification for use outdoors.

When data are to be collected over a prolonged period, calibration facilities must be considered (see Sect. 3). These involve a standard lamp (Appendix B) with dark-room facilities or a suitable housing, and a stabilised power supply as a minimum. Alternatively, calibration services are provided by some of the instrument manufacturers and standards laboratories, either as a visiting standard instrument or by moving the test instrument to the standard facilities. Calibration frequency will depend upon the type of instrument, its mode of operation and operators experience. Trend analysis is only possible when the instrument has proven stability over a long period, with less uncertainty in the consistency of the measurement system than the size of the trend.

Spectral data collection needs further thought about the wavelength range and resolution, the frequency of scans and the acceptable level of straylight and noise. Most scanning spectroradiometers take several minutes to scan across the entire UV range at a resolution of 1 nm. If better resolution or a wavelength range extending into the visible is desirable, scan time increases. Very frequent scanning is therefore not possible. This task is best performed by an instrument with a diode array detector, but at the expense of good straylight rejection, which is important at short UVB wavelengths in the solar spectrum. The higher the specifications placed on an instrument, the higher the price is likely to be.

6.1 Ancillary Data

UV data alone are of limited use. To understand why the observed levels are as they are, what effect they have (Sect. 6.2) and how they may be related to expectations for UV irradiances and effects elsewhere, additional information is necessary. Studies of UV radiation itself benefit from colocated measurements of longer-wavelength or total solar radiation; e.g. the NSF spectroradiometer also measures visible radiation (Booth et al. 1994), which can be used to determine if a day or season is unusually cloudy or the atmosphere turbid, and so helps to confirm whether or not changes in UV radiation are due to changes in ozone or other factors.

Total solar radiation is measured far more widely than UV radiation and can be used to determine the actual or expected amount of clear-sky periods, to estimate atmospheric turbidity, and as a tenuous guide to UV radiation levels at a site where there are no measurements (Webb and Steven 1986).

UV effects studies naturally require information about the effect on the test system of irradiation: for example, use of national cancer registries for skin cancer studies, or some measurement of plant growth for agricultural studies. If plant studies are not in environmentally controlled chambers, then details of temperature, rainfall, etc., are also needed, as multiple stresses can change plant responses to UV radiation (Murali and Teramura 1986; Teramura et al. 1990b).

Where UV measurements are being compared or used alongside models to validate the model or explore radiative transfer processes and influencing factors, a great deal of ancillary data is desirable (Sect. 1.2): ozone amount and profile, temperature profile of the atmosphere, turbidity, type and amount of aerosol, areal albedo, altitude, and that all for clear skies. Such comprehensive information is rarely available, but in general as many additional data as possible should be collected with the UV measurements.

6.2 Effects of UV Radiation

UVB radiation has the shortest wavelengths and highest photon energies naturally available at the earth's surface. Although the number of incident UVB photons is relatively small, their high energy makes them capable of initiating many reactions in both living and inanimate material. Many, but not all, of these reactions lead to undesirable effects.

6.2.1 Human Health

The most researched and publicised aspect of UV radiation effects is erythema (sunburn), which can be associated with skin cancer. There are three types of skin cancer: the unpleasant but infrequently fatal basal and squamous cell carcinomas, and the more lethal malignant melanoma. There is a clear relation between the carcinomas and total lifetime exposure to erythemally effective radiation (UVB and UVA radiation weighted with the erythemal action spectrum: each wavelength's effectiveness at producing sunburn) (Elwood 1989). The incidence of skin cancer has increased over the past couple of decades as travel to lower latitudes, fashion and increased leisure and life expectancy have all led to increased cumulative exposure to sunlight, although other contributory factors besides sunlight cannot be ruled out (Moan and Dahlback 1993). The causes of malignant melanoma are more complex and are discussed by Elwood (1994), but are still associated with exposure to UV radiation.

A more complex and less well documented result of UV irradiation of skin is its effect on a part of the immune system. Much of the experimental work on this subject has been conducted on animal models, but recently some of the same responses have been observed in humans. In mice some cell-mediated immune responses are suppressed by UV radiation, with an action spectrum of maximum response in the UVC, decreasing across the UVB to 3% of the maximum at 320 nm (DeFabo and Noonan 1983). This immune suppression is thought to play a role in the development of non-melanoma skin cancer (DeGruijl and Van der Leuen 1982; Ward and Schreiber 1991) and may help determine the development of other skin and infectious diseases, e.g. herpes and leishmaniasis (Giannini 1986, Yasumoto et al. 1987; Norval et al. 1989) and protect against autoimmune reactions. The immuno-suppressive effect of UV irradiation can be demonstrated and tested for by contact hypersensitivity in animal models (Noonan et al. 1981) and in humans (Hersey et al. 1983; Kalimo et al. 1983), where a dose-dependent suppression has been described (Cooper et al. 1992). Animal experiments have shown that pigmentation does not affect UV immune suppression, although different mouse strains showed different susceptibilities to UV suppression (Noonan and DeFabo 1990). In humans both caucasians and blacks have been immuno-suppressed by UV radiation (Vermeer et al. 1991), following the animal model, and skin cancer patients have shown lower responses to contact hypersensitivity than UV-irradiated controls, suggesting that the cancer patients may be more susceptible to UV suppression (Yoshikawa et al. 1990). Urocanic acid has been postulated as a unique photoreceptor in mammalian skin responsible for the immuno-suppressive effect, and the evidence for this is reviewed by Noonan and DeFabo (1993). UV suppression of the immune system is selective, acting only on certain functions and leaving the rest of the immune system to function normally, but increased UV-induced suppression has the potential to influence health in a number of ways; it may influence susceptibilitiy to natural infectious agents, reduce the effectiveness of some vaccination programmes, increase the reactivation of latent viruses (e.g. herpes) and increase the risk of infection in already immunocompromised people (e.g. those who are HIV-positive), but more work is needed to show whether any of these effects are more than hypothetical.

The effects discussed above have all been detrimental to human health, but UVB radiation does have some positive benefits. The cutaneous synthesis of vitamin D is initiated by the photoconversion of 7-dehydrocholesterol to previtamin D, a reaction that requires irradiation by wavelengths less 316 nm. The previtamin D then undergoes a slow heat isomerisation to vitamin D. In addition to beginning the process of producing vitamin D, UVB (and UVA) radiation allow a finely balanced control of the amount of vitamin D available from cutaneous synthesis by limiting previtamin D accumulation through a photoequilibrium of previtamin D photoproducts, and by degrading excess vitamin D remaining in the skin. Reviews of the photochemistry and photobiology of vitamin D synthesis are given by Webb (1993) and Webb and Holick (1988). Vitamin D plays an essential role in calcium metabolism and is necessary for a healthy skeleton: a lack of the vitamin leads to rickets and osteomalacia. Dietary sources of vitamin D include fatty fish and some fortified foods, e.g. milk in the USA, but a few minutes exposure to sunlight in most parts of the world will provide for an adequate vitamin D status at a suberythemal dose of UVB.

UV radiation is also used to treat some skin disorders, e.g. psoriasis. Treatment often involves irradiation with artificial light sources, sometimes in combination with drugs [e.g. PUVA (psoralen plus UVA) therapy], but sunlight can also have a natural curative effect for these diseases.

The skin is one organ directly exposed to UV radiation; another is the eye. UVB damage to the cornea and the lens have been well documented, but there is no conclusive evidence that solar UVB radiation damages the retina of a normal eye, because the lens absorbs almost all the UVB and UVA radiation (Taylor 1989). The increased prevalence of cataracts in sunny regions and the tropics has long been recognised (e.g. Hiller et al. 1977), and increased exposure to solar UV radiation has been postulated as the reason for this observation (Taylor 1980b, Hallows and Moran 1981). Epidemiological studies (Taylor et al. 1988) have since shown that cortical and posterior subcapsular cataracts are associated with cumulative UVB exposure (but not UVA exposure), and that activities related to sunlight exposure increase the risk of cortical and mixed cataracts (Boschow et al. 1989; Italian-American Study Group 1991). Nuclear cataracts did not depend upon UV exposure in any of these studies. Corneal damage associated with UV exposures includes photokeratitis (snow blindness), pterygium and climatic droplet keratopathy. A number of studies established a clear relation between these diseases and acute (photokeratitis) or cumulative UVB exposure (Taylor 1980a; Moran and Hollows 1984; Taylor 1989), and for pterygium and climatic droplet keratopathy Taylor et al. (1989) an association with UVA exposure was also noted.

6.2.2 Animal Health

All the UV effects on human health can be applied to other mammals as well; indeed, much of the basic research into these effects has been conducted using animals that are similar to the human in some aspect of their anatomy, in the hope that they will respond in the same way: hairless mice are frequently used in skin-based studies, and opossum has also proved a useful model (Ley et al.

1989). Squirrels and rabbits provide models for the human eye (Reddan et al. 1980; Zigman and Paxhia 1988). Like humans, animals evolved to survive in a particular climate, and changing the climate, or the animals' interaction with their environment by domestication, can cause problems.

Observations of domestic animals have shown that squamous cell carcinomas associated with solar radiation are found on the eye and unpigmented lips, nostrils and ears of cattle, horses, cats, dogs and sheep (Madewell and Theilen 1987). Skin lesions (McArdle and Bullock 1987) and cataracts (Allison 1962) attributable to UV radiation have been reported in farmed fish where the fish have little possibility of finding shelter in the shallow ponds in which they are kept. UV radiation also contributes to the incidence of keratoconjunctivitis in cattle (Slatter 1990) and Uberreiter's syndrome in some dogs (Slatter 1990). Economically, any increase in bovine ocular squamous cell carcinoma and keratoconjunctivitis could have serious implications for the beef and dairy industries due to lost productivity and the cost of treatment: breeds susceptible to the diseases are widely bred for beef and milk (Mayer 1992). Similarly, fish farming, which is an increasingly important industry, could be adversely affected. Animal welfare issues may also be raised if UV-related diseases increase: the owners of white cats are already being advised to use sunblock on their cats' ears, especially during the summer months (Mayer 1992).

6.2.3 Aquatic Life

Laboratory studies and investigations of marine life in situ have shown that UV radiation is damaging to a variety of aquatic organisms (reviews by Worrest 1982, 1986; Calkins 1982; Smith 1989; Holm-Hansen et al. 1993; Smith and Cullen 1995). Much of the research has concentrated on phytoplankton, which represent the base of the aquatic food web. UV exposure of phytoplankton reduces growth, photosynthesis, cellular protein and photosynthetic pigments, amongst other deleterious effects. Applying these results to an assessment of UV effects in the natural environment is difficult because of the gradient of UV radiation with depth beneath any water surface (Piazena and Hader 1994) and the changing location of organisms within this gradient either by their own volition or by physical mixing of the water. Even the in situ experiments maintain samples at a fixed depth. Extrapolation of laboratory studies is further complicated by the fact that the motility of organisms is very sensitive to UVB radiation (Hader 1993), and in some cases to UVA. Swimming speed and orientation can be adversely affected but the effects vary from organism to organism. Little is known about the sensing mechanisms for UV radiation, but both visible and long-wavelength UVA are implicated as triggers for phototoxic movement (Foster and Symth 1980). The ability to move would allow an organism in its natural environment protection against overexposure, but only if it senses and responds to UVB before deleterious effects of UVB on motility prevent it from moving.

Studies of other organisms have shown that UV radiation increases mortality of larval crustaceans (Damkaer et al. 1981) and adult planktonic crustaceans (Dey et al. 1988), and reduces fecundity at adult stages of growth, although whether threshold levels for damage are reached in the natural habitat is not conclusively illustrated (Dey et al. 1988). Fish larvae can also suffer severe detrimental effects (Hunter et al. 1982), and coral is visibly damaged (Brown et al. 1994) and killed by varying levels of UV (dependent on species), but once again extrapolating these results to the natural environment is difficult.

6.2.4 Plant Life

The effects of UVB radiation on plants vary widely from species to species, within species, and with respect to other conditions of growth: spectral quality and intensity of radiation (UVB:UVA:PAR), water, CO_2, temperature and other stresses. It is therefore difficult to generalise about UVB effects on plants, or to extrapolate the results of carefully controlled experiments in artificial conditions (including light) to expectations for field-grown crops or natural ecosystems. Conflicting reports of UVB effects on plants, especially in earlier studies, may be attributed to different methods of experimentation. Reviews of current understanding of the results and mechanisms by which UVB radiation influences plants are given by Tevini and Teramura (1989) and Bornman and Teramura (1993).

Plants which are sensitive to UVB radiation often display reductions in growth, which can be measured in a number of ways, e.g. plant height, leaf area, dry weight, yield, etc., although this is not always associated with reduced photosynthesis (Beyschlag et al. 1988). Many plants that reduce leaf area in response to UVB irradiation also show increased leaf thickness (Tevini et al. 1983; Mirecki and Teramura 1984) or a change in leaf composition (Tevini et al. 1983; Britz and Adamse 1994). Higher plants tend to accumulate UV-absorbing pigments in their leaves when irradiated with UVB radiation (Wellmann 1985), and can also increase the outer waxy layer (Steinmuller and Tevini 1985), which may increase reflection of UV radiation (Clark and Lister 1985), but will also reduce transpiration. Reduced transpiration due to stomatal closure has been demonstrated under enhanced UVB for cucumber (Teramura et al. 1983) and sunflowers (Tevini et al. 1988) but not wheat or wild oat (Beyschlag et al. 1988), and this may be one reason for the reduction in photosynthesis which has been observed in growth chambers and field studies of UV-sensitive cultivars of soybean (Biggs et al. 1981; Teramura and Murali 1986), cucumber and sunflower seedlings. Studies with soybeans have shown that the period of growth most affected by UV was the time between vegetative and reproductive stages (Teramura and Sullivan 1987). Pollen is generally well protected from UV radiation, but flowering may be repressed in some species (Caldwell 1968), and if the timing of flowering is affected it may reduce chances of pollination and hence competitiveness of the species.

The number of species and their cultivars tested for UV sensitivity is very small. Amongst the agricultural crops tested, reductions in crop yield under enhanced UVB have been found in squash, tomato, pea, wheat, potato and soybean, but not in rice, peanut, corn or cabbage, lettuce and rape (Biggs and Kossuth 1978; Biggs et al. 1984; Teramura et al. 1990a). In apparent contradiction (Becwar et al. 1982) excluded ambient UVB levels (rather than increasing UVB above ambient, as in the Biggs experiment, and found no effect on the growth of pea, potato, radish or wheat. The work of Teramura and Murali (1986) with 23 soybean cultivars illustrated the range of UVB sensitivity amongst different cultivars. Two of these cultivars, one sensitive and one tolerant to UVB, were studied over six growing seasons under two levels of enhanced UVB (Teramura et al. 1990b). Reduction in yield was greatest in the sensitive cultivar, but was only evident in 4 of the 6 years. During 2 years of drought the UVB effects were masked by growth reductions due to water stress.

In contrast to annual crops, trees can be studied over a prolonged period to observe cumulative effects of UVB radiation and adaptive, protective or repair mechanisms. Excluding ambient UVB from tree seedlings increased growth in some species (Bodenreider and Klein 1982) but had no effect on two high-elevation species of conifer (Kaufman 1978). Of 15 conifer species tested (Kossuth and Biggs 1981; Sullivan and Teramura 1988), 47% showed reductions in dry matter with elevated UVB, while biomass increased in two high-elevation species and was unchanged in others. The Loblolly pine was most responsive to UVB, with effects on growth, photosynthesis and concentrations of UVB-absorbing compounds appearing to be cumulative throughout the growing season (Sullivan and Teramura 1989). Seeds of the Loblolly pine collected from different latitutdes and grown for 3 years under elevated UVB showed differences in sensitivity to UVB amongst the seed sources (indicating genetic variability), though UVB reduced biomass in trees from all sources (Sullivan and Teramura 1992).

The different effects (or not) of UVB radiation among and within plant species, and the interaction of UVB stress with the many other stresses that can affect field-grown or natural vegetation makes it very difficult to predict the effect of a changing UVB climate on agriculture and ecosystems. However, it is likely that the current balance of natural and agricultural ecosystems will be compromised as an additional stress changes competitive advantages between plants.

6.2.5 Weathering of Materials

Materials such as wood, plastics and paint degrade when used for outdoor applications due to the deleterious effects of temperature, moisture and UV radiation. In general, the lifetime of the material is determined by its light stability, although harsh temperature and humidity environments may exacerbate the weathering due to solar radiation (Hamid et al. 1988; UNEP 1989; Amin et al. 1991). To ensure an adequate material lifetime plastic formulations usually include a light stabilizer, and wood is surface-treated or painted to protect it against both light and moisture. Paint and polymer coatings are extensively used elsewhere, e.g. the automotive industry, for both protection from the elements and cosmetic appeal, yet they are themselves subject to weathering. The most damaging wavelengths are generally those in the UV region of the spectrum, especially the UVB (UNEP 1991). The outdoor lifetime of materials is reduced in climates with high UV irradiances, e.g. the tropics, and any increase in UVB due to ozone depletion would reduce the lifetime of materials used in numerous building, transport, agricultural and other applications. Visual evidence of degradation is usually a yellowing of the material, or for coloured coatings a colour change. The discoloration may be irregular and is also an indicator of structural damage, at least in the small number of plastics tested (Decker 1984; Andrady et al. 1989). The serviceable lifetime of polymer materials under high UV irradiances might be increased if higher levels of stabilizer were added during their manufacture, but more research is needed into the practicality of this approach and the stabilizers which could be used. The action spectrum of weathering for many plastic mixes is unknown and research into this area of UV effects has not gained the same attention as research into the health effects, despite the economic implications.

7 Conclusion

There is an increasing demand for solar UV radiation data due to observed stratospheric ozone depletion and the potential consequences for the biosphere, yet the measurement facilities are sparse and the instrumentation (and therefore data types) diverse. Spectroradiometers offer high-resolution spectral data which can be applied to many studies, but require skilled operation and maintenance. Broadband radiometers are simpler and more practical but provide data with limited uses. In each case, careful calibration is both essential and a problem in itself. Dosimeters of various design are available for surface-specific studies, but it is difficult to extrapolate the results to situations and biological responses other than the one represented by the dosimeter. UV radiation climatology can be modelled in a number of ways, but is always dependent on well-specified input data to the model. In clear sky conditions models generally agree with surface radiation measurements, with suitable choice of albedo and aerosol assessments. However, incorporating cloud into the models remains a challenge, and any model estimates will have to be verified by ground-based observations.

A number of national commitments have been made to UV monitoring, with international cooperation in some areas, but there is no worldwide organisation of UV measurements and monitoring sites are unevenly distributed around the globe. A recent initiative by the World Meteorological Organisation may mean that in future UV data may be accessed through a central facility, but only data that are measured can be reported. There is a continuing need to extend and improve the solar UV monitoring facilities at the earth's surface.

Appendix A: List of Instrument Manufacturers

Broadband

Eppley Laboratory Inc.
12 Sheffield Avenue
Newport, Rhode Island 02840
USA

International Light Co.
17 Graf Rd.
Newburyport, Massachusetts 01950
USA

Solar Light Co. Inc.
721 Oak Lane
Philadelphia, Pennsylvania 19126
USA

Vital Technologies Corp.
680 Hardwick Rd. Unit 3
Bolton
Ontario, L7E 5R4
Canada

Yankee Environmental Systems Inc.
101 Industrial Rd.
P.O.Box 746
Turners Falls, Massachusetts 01376
USA

Spectral

Bentham Instruments Ltd.
2 Boulton Road
Reading RG2 ONH
UK

Biospherical Instruments Inc.
5340 Riley St.
San Diego, California 92110-2621
USA

LI-COR Inc.
P.O.Box 4425
Lincoln, Nebraska 68504
USA

Macam Photometrics Ltd.
10 Kelvin Square
Livingston EH 54 5PF
Scotland

Optronic Laboratories Inc.
4470 35th St.
Orlando, Florida 32811
USA

Sci-Tec Instruments Inc. (Brewer)
1526 Fletcher Rd.
Saskatoon
Saskatchewan, S7M 5M1
Canada

Multi-channel Instruments

Biospherical Instruments Inc.
See above right.

Yankee Environmental Systems
See above left.

Appendix B: List of National Standards Laboratories and Other Sources of Standard Lamps

Standards Laboratories

National Physical Laboratories
Teddington
Middlesex
TW11 OLW
UK

National Institute of Standards and Technology
Gaithersburg, Maryland 20899
USA

Physikalisch-Technische Bundesanstalt
Bundesallee 100
38116 Braunschweig
Germany

CSIRO National Measurement Laboratory
Bradfield Rd.
Linfield, NSW
Australia 2070

VNIIOFI (National Standards Laboratory)
Russia

Calibrated Lamps (Traceable to)

Eppley Laboratories (NIST)
See Appendix A

Optronic Laboratories (NIST)
See Appendix A

Appendix C: Glossary of Terms

Radiant flux	Total power emitted by a source (watts, W)
Radiant intensity	Flux emitted per unit solid angle (W/steradian)
Radiance	Flux emitted per unit of observed area $[W/(sr \cdot m^2)]$
Irradiance	Flux density received by a surface (W/m^2)
Cosine response	When measuring irradiance on a horizontal surface the receiving optics should transmit radiation arriving at an angle Θ from the normal in such a way that the measured signal is $\cos \Theta$ x (signal from radiation at normal incidence). The integrated incident radiation from the whole hemisphere treated in this way gives the measured irradiance. In reality foreoptics with a good cosine response are difficult to find, leading to errors in the measurements.
Spectral resolution	The smallest incremental change (in wavelength) that can be resolved and measured.
Sensitivity	The lowest level of irradiance, or change in irradiance, that can be measured.
Straylight	Radiation at wavelengths other than the one(s) being measured that enters the detector and is attributed to the nominated wavelength(s). Straylight can result from the instrument's slit function (pass band), or from misguided light within the instrument.
Slit function	The shape of a single monochromatic line measured by a spectrometer as it scans across a waveband centred on that line. Ideally a single line, the slit function is often approximately triangular in shape with the apex at the wavelength of the incident radiation.
FWHM	Full width half maximum: the width (in nm) of the slit function at half the height (measured irradiance) of the peak value of the slit function.
Pass band	Similar to the slit function. The (narrow) waveband over which a channel of a multi-filter instrument measures irradiance. Ideally a rectangular step function, in reality often a broad triangle which is frequently asymmetric.
Spectral response	The function which describes the sensitivity of a (broadband) instrument to each wavelength of incident radiation. It is often designed to approximate a biological action spectrum e.g. that for erythema.
Action spectrum	The spectrum showing the effectiveness of each wavelength in producing a given reaction.

References

Allison LN (1962) Cataract among hatchery reared lake trout. Prog Fish Cult 22:155

Amin MB, Hamid SH, Maadhah AG, Al-Jarallah AM, Abu-Abdoun IT (1991) Weather induced degradation of plastic products in Saudi Arabia. KACST Project No AR-10-052, Technical progress report no 4

Andrady AL, Torikai A, Fueki K (1989) Photodegradation of rigid PVC formulations I. Wavelength sensitivity to light-induced yellowing by monochromatic light. J Appl Polym Sci 37:935–946

Austin J, Barwell BR, Cox SJ, Hughes PA, Knight JR, Ross G, Sinclair P, Webb AR (1994) The diagnosis and forecast of clear sky ultraviolet levels at the earth' surface. Metereol Applic 1:321–336

Bais AF, Zerefos CS, Meleti C, Ziomass IC, Tourpali K (1993) Spectral measurements of solar UVB radiation and its relation to total ozone, SO_2 and cloud. J Geophys Res 98:5199–5204

Barton IJ, Paltridge GW (1979) The Australian climatology of biologically effective ultraviolet radiation. Aust J Dermatol 20:68

Becwar MR, Moore FD, Burke MJ (1982) Effects of depletion and enhancement of ultraviolet-B (280–315 nm) radiation on plants grown at 3000 m elevation. J Am Soc Hortic Sci 107:771–774

Bener P (1960) Investigation of the spectral intensity of ultraviolet sky and sun+sky radiation (between 297.5 and 370 µm) under different conditions of cloudless weather at 1590 masl. Contract AF61(052)–54, Technical summary report no 1, Physikalisch-Meteorologisches Observatorium Davos, Davos Platz, Switzerland

Bener P (1963) The diurnal and annual variations of the spectral intensity of sky and global radiation on cloudless days at Davos, 1 590 masl. Contract AF61(052)-618, Technical summary report no 2 Physikalisch-Meteorologisches Observatorium Davos, Davos Platz, Switzerland

Bener P (1964) Investigation of the effect of clouds on ultraviolet sky radiation (at 330 to 370 µm). Contract AF61(052)-618, Technical summary report no 2 Physikalisch-Meteorologisches Observatorium Davos, Davos Platz, Switzerland

Bener P (1969) Spectral intensity of natural UVR and its dependence on various parameters. In: Urbach F (ed) The biological effects of UVR. Pergamon Press, Oxford

Berger DS (1976) The sunburning ultraviolet meter: design and performance. Photochem Photobiol 24:587–583

Beyschlag W, Barnes PW, Lint SD, Caldwell MM (1988) Enhanced UVB radiation has no effect on photosynthetic characteristics of wheat and wild oat under greenhouse and field conditions. Photosynthetica 22(5):516–525

Biggs RH, Kossuth SV (1978) Effects of ultraviolet-B radiation enhancements under field conditions on potatoes, tomatoes, corn, rice, southern peas, peanuts, squash, mustard and radish. In: UVB biological and climatic effects research (BACER), Final report, Environment Protection Agency (EPA), Washington, DC

Biggs RH, Kossuth SV, Teramura AH (1981) Response of nineteen cultivars of soybeans to ultraviolet-B irradiance. Physiol Plant 53:19–26

Biggs, RH, Webb PG, Garrard LA, Sinclair TR, West SH (1984) The effects of enhanced ultraviolet-B radiation on rice, wheat, corn, soybean, citrus and duckweed. Year 3 interim report. EPA (Environmental Protection Agency) Rep 808075-03, Washington DC

Blumthaler M, Ambach W (1990) Indication of increasing ultraviolet-B radiation in alpine regions. Science 248:206–208

Blumthaler M, Ambach W, Roland S, Staehelin J (1994a) Erythemal UV-B irradiance (Robertson-Berger meter sunbrun data) under ozone deficiencies in winter/spring 1993. Photoechem Photobiol 59:657–659

Blumthaler M, Webb AR, Seckmeyer G, Bais AF, Huber M, Mayer B (1994b) Simultaneous spectroradiometry: a study of solar UV irradiance at two altitudes. Geophys Res Lett 21(25):2805–2808

Bogenreider A, Klein R (1982) Does solar UV influence the competitive relationship in higher plants? In: Calkins J (ed) The role of solar ultraviolet radiation in marine ecosystems. Plenum, New York, pp 641–649

Booth RC, Lucas T, Moorow J, Weiler S, Penhale P (1994) The United States National Science Foundation/Antarctic Program's network for monitoring ultraviolet radiation. In: Penhale P, Weiler S (eds) Ultraviolet radiation and biological research in Antarctica, AGU Antarctic Research Series. AGU, Washington, DC

Bornman JF, Teramura AH (1993) Effects of ultraviolet-B radiation on terrestrial plants. In: Young AR, Bjorn LO, Moan J, Nultsch W (eds) Environmental UV photobiology. Plenum, New York, chap 14

Boschow TR, West SK, Azar A, Munoz B, Sommer A, Taylor HR (1989) Ultraviolet light exposure and risk of posterior subcapsular cataracts. Arch Opthalmol 107:369–372

Brewer AW (1973) A replacement for the Dobson spetrophotometer. Pure Appl. Geophys 106–8, V–VIII:919–927

Britz ST, Adamse P (1994) UV-B induced increase in specific leaf weight of cucumber as a consequence of increased starch content. Photochem Photobiol 60(2):116–119

Brown BE, Dunne RP, Scoffin TP, Le Tissier MDA (1994) Solar damage in intertidal corals. Mar Ecol Prog Ser 105:219–230

Bruhl C, Crutzen PJ (1989) On the disproportionate role of tropospheric ozone as a filter against solar UVB radiation. Geophys Res Lett 16:703–706

Caldwell MM (1968) Solar ultraviolet irradiation as an ecological factor for alpine plants. Ecol Monogr 38:243–268

Calkins JC (1982) The role of solar ultraviolet light in marine ecosystems. Plenum, New York

CIAP (1974) In: Grobecker AJ, Coroniti SC, Cannon RH (eds) The effects of stratospheric pollution by aircraft. US Department of Transportation, Washington, DC

Clark JB, Lister GR (1975) Photosynthetic action spectra of trees. II. The relationship of cuticle structure to the visible and ultaviolet spectral properties of needles from four coniferous species. Plant Physiol 55:407–413

COMESA (1975) Report of the committee on meteorological effects of stratospheric aircraft. Meteorological Office, Bracknell, UK

Cooper KD, Oberhelman L, Hamilton TA, Baadsgaard O, Terhune M, Levee G, Anderson T, Koren H (1992) UV exposure reduces immunisation rates and promotes tolerance to epicutaneous in humans – relationship to dose CD1a-DR+ epidermal macrophage induction and Langerhans cell depletion. Proc Natl Acad Sci U.S.A. 89:8497–8501

Correll DL, Clark CO, Goldberg B, Goodrich VR, Hayes DR, Klein WH, Schere WD (1992) Spectral ultraviolet-B radiation fluxes at the earth's surface: long-term variations at 39°N, 77°W. J Geophys Res 97:7579–7591

Damkaer D, Dey DB, Heron GA (1981) Dose/dose rate responses of shrimp larvae to UV-B radiation. Oecologia (Berl.) 48:178–182

Davis A, Deane GHW, Diffey BL (1976) Possible dosimeter for ultraviolet radiation. Nature 261:169–170

Decker C (1984) In: Owen ED (ed) Degradation and stabilization of PVC. Elsevier Applied Science, New York

DeFabo EC, Noonan FP (1983) Mechanism of immune suppression by ultraviolet radiation in vio. I. Evidence for the existence of a unique photoreceptor in skin and its role in photoimmunology. J Exp Med 158:84–98

DeGruijl FR, Van der Leuen (1982) Systemic influence of preirradiation of a limited skin area on UV tumorigenesis. Photochem Photobiol 35:379–383

DeLuisi JJ, Wendel J, Kreiner F (1992) An examination of the spectral response characteristics of seven Robertson-Berger meters after long-term field use. Photochem Photobiol 56:115–122

Dey DB, Damkaer DM, Heron GA (1988) UV-B dose/dose rate responses of seasonally abundant copepods of Puget Sound. Oecologia 76:321–329

Diffey BL, Kerwin M, Davis A (1977) The anatomical distribution of sunlight. Br J Dermatol 97:407–409

Dobson GMB (1931) A photoelectric spectrophotometer for measuring the amount of atmospheric ozone. Phys Soc XLIII,3:324–339
Elwood JM (1989) The epidemiology of melanoma: its relationship to ultraviolet radiation and ozone depletion. In: Transactions of the Menzies foundation, vol 15. The Menzies Foundation, Melbourne, pp 95–107
Elwood JM (1994) Human melanoma and ultraviolet radiation. In: Stanhill G (ed) Advances in bioclimatology 3. Springer, Berlin Heidelberg New York, pp 1–39
Farman JC, Gardiner BG, Shanklin JD (1985) Large losses of total ozone in Antarctica reveal seasonal CLO NO interaction. Nature 315:207–210
Farmer CB, Toon GC, Schaper PW, Blavier JF, Lowes LL (1987) Stratospheric trace gases in the spring 1986 Antarctic atmosphere. Nature 329:126–128
Feister U, Grewe R (1995) Spectral albedo measurements in the UV and visible region over different types of surface. Photochem Photobiol 62(4):736–744
Forster PM, de F (1994) Measuring and modelling UV radiation. PhD Thesis. Reading University, Reading, UK
Foster KW, Smyth RD (1980) Light antennas in phototactic algae. Microbiol Rev 44:572–630
Frederick JE, Weatherhead EC (1992) Temporal changes in surface ultraviolet radiation: a study of Robertson-Berger meter and Dobson data records. Photochem Photobiol 56:123–131
Frederick JE, Diaz SB, Smolskaia I, Esposito W, Lucas T, Booth CR (1994) Ultraviolet solar radiation in the high latitudes of South America. Photochem Photobiol 60:356–362
Garadzha MP, Nezval YE (1987) Proc Symp Climate and Human Health, Leningrad WCAP (World Climate Applications Programme) Report No 2
Gardiner BG, Kirsch PJ (1992) First European intercomparison of ultraviolet spectroradiometers. CED (Commission of European Communities) Air Pollution Report 38, Brussels
Gardiner BG, Kirsch PJ (1994) Second European intercomparison of ultraviolet spectroradiometers. CEC (Commission of European Communities) Air Pollution Report 49, Brussels
Gardiner BG, Kirsch PJ (1995) Setting Standards for European ultraviolet spectroradiometers. CEC (Commission of European Communities) Air Pollution Report 53, Brussels
Gardiner BG, Webb AR, Bais AF, Blumthaler M, Dirmhirn I et al. (1993) European intercomparison of ultraviolet spectroradiometers. Environ Technol 14(1):25–43
Giannini MSH (1986) Suppression of pathogenesis in cutaneous leishmaniasis by UV irradiation. Infect Immun 51:838–843
Gleason JF, Bhartia PK, Herman JR et al. (1993) Record low global ozone in 1992. Science 260:523–526
Green AES, Cross KR, Smith LA (1980) Improved analytic characterisation of ultraviolet skylight. Photochem Photobiol 31:59–65
Hader, DP (1993) Risks of enhanced solar ultraviolet radiation for aquatic ecosystems. Prog Phycol Res 9:1–45
Hamid SH, Maadha HG, Qureshi FS, Amin MB (1988) Weather induced degradation of polyethylene: Infrared spectroscopic study. Arabian J Sci Eng 13:503
Herlihy E, Gies PH, Roy CR, Jones M (1994) Personal dosimetry of solar ultraviolet radiation for different outdoor activities. Photochem Photobiol 60:288–294
Herman JR, Larko D (1994) Low ozone amounts during 1992–1993 from Nimbus 7 and Meteor 3 total ozone mapping spectrometers. Geophys Res Lett 99D2:3483–3496
Hersey P, Hasic E, Edwards A, Bradley M, Haran G (1983) Immunological effects of solarium exposure. Lancet I:545:548
Hiller R, Giacometti L, Yuen K (1977) Sunlight and cataract: an epidemiological investigation. Am J Epidemiol 105:450–459
Hisdal V (1986) Spectral distribution of global and diffuse solar radiation in Ny-Alesund, Spitsbergen. Polar Res 5:1–27
Hollows R, Moran D (1981) Cataract – the ultraviolet risk factor. Lancet II:1249–1250
Holm-Hansen O, Lubin D, Helbling EW (1993) Ultraviolet radiation and its effects on organisms in the aquatic environment. In: Young AR, Bjorn LO, Moan J, Nultsch W (eds) Environmental UV photobiology. Plenum, New York, chap 13

Hunter JR, Kaup SE, Taylor JH (1982) Assessments of effects of UV radiation on marine fish larvae. In: Calkins J (ed) The role of solair ultraviolet radiation in marine ecosystems. Plenum, New York, pp 459–497

Italian-American Cataract Study Group (1991) Risk factors for age related cortical, nuclear and posterior subcapsular cataracts. Am J Epidemiol 133, 541–553

Ito T, Sakoda Y, Uekubo T, Naganuma H, Fukuda M, Hayashi M (1993) Scientific application of solar UVB observations from JMA network. Proc 13th UOEH Symp, Kitakyushu, Japan, Oct 1993

Justus CG, Murphy BB (1994) Temporal trends in surface irradiance at ultraviolet wavelengths. J Geophys Res 99D1:1389–1394

Kalimo K, Koulu L, Jansen CT (1983) Effect of a single UVB or PUVA exposure on immediate and delayed skin hypersensitivity reactions in humans. Arch Dermatol Res 275:374–378

Kaufman MR (1978) The effect of ultraviolet (UVB) radiation on Engelmann spruce and lodgepole pine seedlings. In: EPA-IAG-D6-0168 BACER Program EPA (Environmental Protection Agency), Washington, DC

Kennedy BC, Sharp WE (1992) A validation study of the Robertson-Berger meter. Photochem Photobiol 56:133–141

Kerr JB, McElroy CT (1993) Evidence for large upward trends of ultraviolet-B radiation linked to ozone depletion. Science 262:1032–1034

Kerr JB, McElroy CT, Tarasick DW, Wardle DI (1994) The Canadian Ozone Watch and UV-B advisory programs. Proc 1992 Quad Ozone Symp, Chalottesville, Virginia

Komhyr WD, Grass RD, Evans RD, Leonard RK, Quincy DM (1994) Unprecedented 1993 ozone decrease over the Unites States from Dobson spectrophotometer observations. Geophys Res Lett 21:201–204

Koskela T (ed) (1994) The Nordic intercomparison of ultraviolet and total ozone instruments at Izana from 24 October to 5 November 1993. Final rep, Finnish Meteorological Institute, Helsinki

Kossuth SV, Biggs RH (1981) Ultraviolet-B radiation effects on early seedling growth of *Pinaceae* species. Can J For Res 11, 243–248

Kremser H, Koepke P, Quenzel H (1984) Aerosol optical thickness from direct solar radiation: improved Langley method applied to measured data. In: Fiocco G (ed) Current problems in atmospheric radiation. Deepack, Hampton, Virginia, pp 546–549

Leszczynski K, Jikela K, Ylianttila L, Visura R, Blumthaler M (1997) Report of the WMO/STUK intercomparison of erythemally weighted solar UV radiometers. Rep no 12 Environmental Pollution Monitoring and Research Programme. WMO-TD no 781, WMO, Geneva

Ley RD, Applegate LA, Padialla LS, Stuart TD (1989) Ultraviolet radiation induced malignant melanoma in *Monodelphis domestica.* Photochem Photobiol 50:1–5

Liu SC, McKeen SA, Madronich S (1991) Effect of anthropogenic aerosols on biologically active ultraviolet radiation. Geophys Res Letts 18,12:2265–2268

Lubin D, Frederick JE (1989) Measurements of enhanced springtime ultraviolet radiation at Palmer Station, Antarctica. Geophys Res Lett 16:783–785

McLaughlin JA, Anderson RR, Holick MF (1982) Spectral character of sunlight modulates phytosynthesis of previtamin D3 and its photoisomers in human skin. Science 216:1001–1003

Madewell BR, Theilen GH (eds) (1987) Veterinary cancer medicine. Lea and Febiger, Philadelphia

Madronich S (1992) Implications of recent total atmospheric ozone measurements for biologically active ultraviolet radiation reaching the earth's surface. Geophys Res Lett 19:37–40

Madronich S (1993) The atmosphere and UVB radiation at ground level. In: Bjorn LD, Moan J, Nultsch W, Young AR (eds) Environmental UV photobiology. Plenum, New York

Madronich S, DeGruijl F (1994) Stratospheric ozone depletion between 1979 and 1992: implications for biologically active ultraviolet B radiation and non-melanoma skin cancer incidence. Photochem Photobiol 59:541–546

Mayer SJ (1992) Stratospheric ozone depletion and animal health. Vet Rec 131:120–122

McArdle J, Bullock AM (1987) Solar ultraviolet radiation as a causal factor of "summer syndrome" in cage reared Atlantic salmon, *Salmo solar;* a clinical and histopathological study. J Fish Dis 10:255–264
McKenzie RL, Johnston PV, Kotkamp M, Bittar A, Hamlin JD (1992) Solar ultraviolet spectroradiometry in New Zealand: instrumentation and sample results from 1990. Appl Optics 31(30):6501–6509
McKenzie RL, Kotkamp M, Seckmeyer G, Erb R, Roy CR, Gies, HP, Toomey SJ (1993) First southern hemisphere intercomparison of solar UV spectra. Geophys Res Lett 20,20:2223–2226
McKinlay AF, Diffey BL (1987) A reference action spectrum for ultraviolet induced erythema in human skin. CIE J 6:17
Michelangeli DV, Allen M, Young YL, Shia RL, Crisp D, Eluszkiewicz J (1992) Enhancement of atmospheric radiation by an atmospheric aerosol layer. J Geophys Res 97, D1, 865–874
Mirecki RM, Teramura RH (1984) Effects of ultraviolet-B irradiance on soybean. V. The dependence of plant sensitivity on the photosynthetic photon flux density during and after leaf expansion. Plant Physiol 74:475–480
Moan J, Dahlback A (1993) Ultraviolet radiation and skin cancer: epidemiological data from Scandinavia. In: Young AR, Bjorn Lo, Moan J, Nultsch W (eds) Environmental UV photobiology. Plenum, New York
Molina M, Rowland FS (1974) Stratospheric sink for chlorofluormethanes - chlorine atom catalyzed destruction of ozone. Nature 249:810–812
Moran J, Hollows FC (1984) Pterygium and ultraviolet radiation : a positive correlation. Br J Ophthalmol 68:343–346
Murali NS, Teramura AH (1986) Effectiveness of UV-B radiation on the growth and physiology of field-grown soybean modified by water stress. Photochem Photobiol 44:215–219
Nicolet M (1989) Solar spectral irradiances with their diversity between 120 and 900 nm. Planet Space Sci 37(18):1249–1289
Noonan FC, DeFabo EC (1990) Ultraviolet-B dose response curves for local and systemic immunosuppression are identical. Photochem Photobiol 52:801–810
Noonan FC, DeFabo EC (1993) UV Induced immunosuppression. In: Young AR, Bjorn LO, Moan J, Nultsch W (eds) Environmental UV photobiology. Plenum, New York, chap 5
Noonan FC, DeFabo EC (1981) Suppression of contact hypersensitivity by ultraviolet radiation: an experimental model. Springer Semin Immunopathol 4:293–304
Norval M, Simpson TJ, Ross JA (1989) Urocanic acid and immunosuppression. Photochem Photobiol 50:267–275
Nunez M, Kalma JD (1995) Satellite mapping of the surface radiation budget. In: Stanhill G (ed) Advances in bioclimatology. Springer, Berlin Heidelberg New York, pp 63–124
Piazena H, Hader DP (1994) Penetration of solar UV irradiation in coastal lagoons of the Southern Baltic Sea and its effect on phytoplankton communities. Photochem Photobiol 60(5):463–470
Reddan JR, Friedman TB, Mostafapor MK (1980) Establishment of epithelial cell lenses from individual rabbit lenses. J Tissue Cult Methods 6:57–60
Ronto G, Gaspar, Grof P, Berces A, Gugolya Z (1994) Ultraviolet dosimetry in outdoor measurements based on bacteriophage T7 as a biosensor. Photochem Photobiol 59:209–214
Schippnick PF, Green AES (1982) Analytical characterisation of spectral actinic flux and spectral irradiance in the middle ultraviolet. Photochem Photobiol 35:89–89
WMO (1995) Scientific Assessment of Ozone Depletion: 1994. WMO Global Ozone Research and Monitoring Project, Report 37. WMO, Geneva
Scotto J, Cotton G, Urbach F, Berger D, Fears T (1988) Biologically effective ultraviolet radiation: measurements in the United States 1974–1985. Science 239:762–764
Seckmeyer G, Bernhard G (1993) Cosine correction of spectral UV irradiances. In: Stamnes K (ed) Atmospheric radiation. Proc SPIE 2049:140–151
Seckmeyer G, McKenzie RL (1992) Increased ultraviolet radiation in New Zealand (45 S) relative to Germany (48 N). Nature 359:135–137

Seckmeyer G, Mayer B, Erb R, Bernhard G (1994) UVB in Germany higher in 1993 than in 1992. Geophys Res Lett 21(7):577–580
Seckmeyer G, Mayer B, Bernhard G, McKenzie RL et al. (1995) Geographical differences in the UV measured by intercompared spectroradiometers. Geophys Res Lett 22:1889–1892
Slatter D (1990) Fundamentals of veterinary ophthalmology, 2nd edn. W. B. Saunders, Philadelphia
Smith JG, Ryan KG (1993) The effect of changes or differences in Robertson-Berger radiometer responsitivity on solar ultraviolet-B measurement. Photochem Photobiol 58:512–514
Smith RC (1989) Ozone, Middle ultraviolet radiation and the aquatic environment. Photochem Photobiol 50:459–468
Smith RC, Cullen JJ (1995) Implications of increased solar UVB for aquatic ecosystems. US National Report to the IUGG (1991–1994) American Geophysical Union, Washington, DC
SORG (1993) Stratospheric ozone 1993. United Kingdom Stratospheric Ozone Review Group. HMSO, London
Stamnes K, Tsay SC, Wiscombe WJ, Jayaweera K (1988) Numerically stable algorithm for discrete ordinate method transfer in multiple scattering an emitting layer media. Appl. Optics 27:2502–2509
Stamnes K, Slusser J, Bowen M (1991) Derivation of total ozone abundance, and cloud effects from spectral irradiance measurements. Appl Optics 30:4418–4426
Steinmuller D, Tevini M (1985) Action of ultraviolet radiation (UV-B) upon cuticular waxes in some crop plants. Planta 164:557–564
Stolarski RS, Bloomfield P, McPeters RD, Herman JR (1991) Total ozone trends deduced from Nimbus 7 TOMS data. Geophys Res Lett 18(6):1015–1018
Stowe LL, Carey RM, Pellegrino PP (1992) Monitoring the Mount Pinatubo aerosol layer with NOAA/11 AVHRR data. Geophys Res Lett 19:159–162
Sullivan JH, Teramura AH (1988) Effects of ultraviolet-B irradiation of seedling growth in the pinaceae. Am J Bot 75:225–230
Sullivan JH, Teramura AH (1989) The effects of ultraviolet-B radiation on Loblolly pines I. Growth, photosynthesis and pigment production in greenhouse grown saplings. Physiol Plant 77:202–207
Sullivan JH, Teramura AH (1992) The effects of ultraviolet-B radiation on Loblolly pines 2. Growth of field grown seedlings. Tree Struct Funct 6:115–120
Tarrant AWS, Brock JR (1975) Further studies of the spectral power distribution of daylight in the ultraviolet region. 18th Session CIE, Publ 36, London
Taylor HR (1980a) Aetiology of climatic droplet keratopothy and pterygium. Br J Ophthalmol 64:154–163
Taylor HR (1980b) The environment and the lens. Br J Ophthalmol 64:303–310
Taylor HR (1989) The biological effects of UVB on the eye. Photochem Photobiol 50:489–492
Taylor HR, West K, Rosenthal FS, Munoz B, Newland HS, Abbey H, Emmett EA (1988) Effect of ultraviolet radiation on cataract formation. N Engl J Med 319:1429–1433
Taylor HR, West K, Rosenthal FS, Munoz B, Newland HS, Emmett EA (1989) Corneal changes associated with chronic UV irradiation. Arch Ophthalmol 107:1481–1484
Teramura AH, Murali NS (1986) Intraspecific differences in growth and yield of soybean exposed to ultraviolet-B radiation under greenhouse and field conditions. Environ Exp Biol 26:89–95
Teramura AH, Sullivan JH (1987) Soybean growth responses to enhanced levels of ultraviolet-B radiation under greenhouse conditions. Am J Bot 74:975–979
Teramura AH, Tevini M, Iwanzik W (1983) Effects of ultraviolet-B irradiation on plants during mild water stress I. Effects on diurnal stomatal resistance. Physiol Plant 57:175–180
Teramura AH, Sullivan JH Ziska LH (1990a) Interaction of elevated ultraviolet-B radiation and CO_2 on productivity and photosynthetic characteristics in wheat, rice and soybean. Plant Physiol 94:470–475
Teramura AH, Sullivan JH, Lydon J (1990b) Effects of UVB radiation on soybean yield and seed quality: a six year field study. Physiol Plant 80:5–11

Tevini M, Teramura AH (1989) UV-B effects on terrestrial plants. Photochem Photobiol 50:479–488
Tevini M, Iwanzik W, Teramura AH (1983) Effects of ultraviolet-B radiation on plants during mild water stress. II. Effects on growth, protein and flavenoid content. Z Pflanzenphysiol 110:459–467
Tevini M, Grusemann P, Fieser G (1988) Assessment of UVB stress by chlorophyll fluorescence analysis. In: Lichtentaler HK (ed) Applications of chlorophyll fluorescence. Kluwer, Dordrecht
Tyrrell RM (1978) Solar dosimetry with repair deficient bacterial spores. Action spectra, photoproduct measurements and a comparison with other biological systems. Photochem Photobiol 27:571–579
UNEP (1989) Environmental effects panel report. UNEP (United Nations Environment Program), Nairobi
UNEP (1991) Environmental effects of ozone depletion: 1991 update. UNEP, Nairobi
Vermeer M, Schneider GJ, Yoshikawa T, Van den Berg JW, Metzman MS, Taylor JR, Streilen JW (1991) Effects of ultraviolet B light on cutaneous immune responses of humans with deeply pigmented skin. J Invest Dermatol 97:729–734; 1032–1034
Walker JH, Saunders RD, Jackson JK, MacSparron DA (1987) Spectral Irradiance Calibrations. National Bureau of Standards, Washington, DC. Spec Publ 250–20
Walker JH, Saunders RG, Jackson JK, Mielenz KD (1991) Results of a CCPR intercomparison of spectral irradiance measurements by national laboratories. J Res Nat Inst Stand Technol 96:647–668
Wang P, Lenoble J (1994) Comparison between measurements and modelling of UV-B irradiance for clear sky – a case study. Appl Optics 33(18):3964–3971
Ward PL, Schreiber H (1991) MHC class I restricted T cells and immune surveillance against transplanted UV light induced tumours. Semin Cancer Biol 2:321–328
Webb AR (1985) Solar UVB ration and vitamin D synthesis in man. Ph D Thesis, Nottingham University, Nottingham, UK
Webb AR (1991) Solar ultraviolet radiation in south-east England: the case for spectral measurements. Photochem Photobiol 54:789–794
Webb AR (1992) Spectral measurements of solar ultraviolet radiation in south-east England. J Appl Meteorol 31:212–217
Webb AR 1996) Historical perspectives of UV monitoring. In: Diffey BL (ed) Measurements and trends in terrestrial UVB radiation in Europe. OEMF, Italy, pp 9–19
Webb AR (1993) Vitamin D synthesis under changing UV spectra. In: Young, AR, Bjorn LO, Moan J, Nultsch W (eds) Environmental UV photobiology. Plenum, New York, chap 7
Webb AR, Campbell GS (1987) Clear sky ultraviolet-B radiation in the English Midlands: a comparison of measured and modelled values. Theor Appl Climatol 38:79–84
Webb AR, Holick MF (1988) The role of sunlight in the cutaneous production of vitamin D3. In: Olson RE (ed) Annual review of nutrition. Annual Reviews Inc, chap 18, Palo Alto, California
Webb AR, Steven MD (1986) Daily totals of solar UVB radiation estimated from routine meteorological measurements. J Climatol 6(4):405–411
Webb AR, Steven MD (1987) Solar ultraviolet-B radiation under cloudless skies. Q J R Meteorol Soc 113:393–400
Webb AR, Weatherhead EC (1997) Current status of UV monitoring. In: Zerefos CS, Bais AF (eds) Solar ultraviolet radiation. Proc NATO ASI, Greece 1995, Springer, Berlin Heidelberg New York
Webb AR, Kline LW, Holick MF (1988) Influence of season and latitude on the cutaneous synthesis of vitamin D3: exposure to winter sunlight in Boston and Edmonton will not promote vitamin D synthesis in human skin. J Clin Endocrinol Metabol 68:882–887
Webb AR, Pilbeam C, Hanafin N, Holick MF (1990) A long-term study to evaluate the relative contributions of exposure to sunlight and diet on the circulating concentrations of 25-OH-D in an elderly population in Boston. Am J Clin Nutr 51:1075–1081

Webb AR, Gardiner BG, Blumthaler M, Forster P, Huber M, Kirsch PJ (1994) A laboratory investigation of two ultraviolet spectroradiometers. Photochem Photobiol 60:84–90

Wellmann E (1985) UV-B Signal/Response-Beziehungen unter natürlichen und artifiziellen Lichtbedingungen. Ber. Dtsch Bot Ges 98:99–104

Wester U (1984) Solar ultraviolet radiation in Stockholm – examples of spectral measurements and influences of measurement error parameters. Internal report RI 1984-03, Dept. Radiation Physics, Karolinska Inst, Stockholm

WMO (1988) Report of the International Ozone Trends Panel 1988. Global Ozone Research and Monitoring Project, Rep no 18, WMO, Geneva

Worrest RC (1982) Review of literature concerning the impact of ultraviolet-B radiation upon marine organisms. In: Calkins J (ed) The role of solar ultraviolet radiation in marine ecosystems. Plenum, New York, pp 429–457

Worrest RC (1986) The effect of solar ultraviolet-B radiation on aquatic systems: an overview. In: Titus JG (ed) Effects of changes in stratospheric ozone and global climate. Overview. EPA and UNEP 1:175–191

Yasumoto S, Hayashi Y, Aurelian L (1987) Immunity to Herpes simplex virus type 2. Suppression of virus induced immune responses in ultraviolet-B irradiated mice. J Immunol 139:2788–2793

Yoshikawa T, Rae V, Bruins-Slot W, Van den Berg J-W, Taylor R, Streilen JW (1990) Susceptibilitiy to the effects of UVB radiation on induction of contact hypersensitivity as a risk factor for skin cancer in humans. J Invest Dermatol 95:530–536

Zheng X, Basher RE (1993) Homogenisation and trend detection analysis of broken series of solar UV-B data. Theor Appl Climatol 47:189–303

Zigman S, Paxhia T (1988) The nature and properties of squirrel lens yellow pigment. Exp Eye Res 47:819–829

Acknowledgements. This work was partially supported by UK Department of the Environment PECD 7/12/18.

Thermal Adaptation and Variable Indoor Climate Control

A. Auliciems[1] and R. de Dear[2]

1 Introduction

Since most of contemporary human living takes place in built environments, the seemingly mundane questions of thermal comfort and moderate thermal stress probably influence as much as 90% of the quality of the human life span. Some sommentators have even gone as fas as hailing thermal comfort as one of the finest achievements of modern civilization (Benzinger 1978).

The role of thermal comfort, and discomfort, in human affairs has been a focus of scholars since Hippocrates of Cos wrote his much cited *Airs and waters* (Jones 1923). Huntington (1924, 1926), and many to follow (Toynbee 1945; Mills 1946; Markham 1947; Missenard 1957; Taylor 1959), variously interpreted the quest for comfort as presenting a double-edged sword to the human condition: it became a major stimulus to human cultural and technological advancement, or, in case of failure, the reason for a group's or even a civilization's decline.

Not surprisingly, periodic reappraisals of this contentious question of climatic determinism appear to go hand in hand with changes to the environmental paradigm (e.g., Sargent and Tromp 1964; Chappell 1975; Lamb 1982), but, less controversially, there can be little argument that the course of human evolution and distribution on the earth's surface may be charted by the organism's responses to prevailing thermal environments. Physiological adaptation and technological development have played complementary roles in permitting human settlement across disparate climate zones well beyond the original, probably warm, human birthplace (Newman 1953, 1955; Burton and Edholm 1955; Scholander 1955, 1956; Sargent 1963, Dubos 1980).

Irrespective of these philosophical questions concerning climate's influence on society, even today's affluent Western society cannot be considered to have completely eradicated the miseries of thermal stress, despite the advances in building design and an abundance of fuel for its amelioration. Indoor climatic surveys in Great Britain (Fox et al. 1973), Belgium, and France (Collins 1979) still find the less privileged subpopulations in society, particularly the elderly with impaired thermoregulatory function, suffering from cold in poorly insulated and under- or even unheated houses. Even relatively affluent North America has been reminded recently of the thermal vulnerability of segments

[1] The University of Queensland, Brisbane, QLD 4072 Australia.
[2] Macquarie University, Sydney, NSW 2109 Australia.

of society, with several hundred elderly citizens dying during the summer of 1995 heat waves in Chicago and elsewhere. Not since the price of fuel oils increased sharply during the "crisis" winter of 1973–1974 have Americans been so directly confronted with the thermal fragility of their society. Furthermore, with the aging populations in economically developed countries such as those of North America and elsewhere in the OECD grouping, direct thermal climatic impacts on society seem certain to increase in magnitude rather than decrease as we move into the twenty-first century.

While thermal comfort is taken for granted by many urban dwellers, its attainment through conversion of energies external to the human body must be regarded with some concern, particularly in view of the environmental impacts resulting from combustion of fossil fuels. For example, about 38% of the USA's annual primary energy is consumed within buildings, with heating and cooling standing out among the largest specific end-uses (LBL 1995).

Therefore, the greenhouse gas emissions directly attributable to our addiction to thermal comfort in the built environment should be more than a little disquieting. Moreover, like the problem of the increasing population of thermally vulnerable elderly citizens, the magnitude of the greenhouse impact of building energy consumption also seems certain to increase rather than decrease in the future. This is due to the sheer scale of the populations in those less developed economies of the world fast approaching their economic "take-off" velocities, which, when attained, will probably see the energy-intensive Western model emulated.

With these post-industrial transformations come seemingly inexorable escalations in aggregate energy consumption and also the percentage of primary energy demand attributable to the pursuit of thermal comfort within built environments. In a rapidly changing world, the long-term well-being of humankind demands that all available options for the management of indoor climates are examined. This should, therefore, include a questioning of some very entrenched approaches along with a reexamination of some basic concepts of building design, and what actually constitutes this nebulous state of mind we refer to as "thermal comfort".

2 The Biophysics and Thermophysiology of Comfort

2.1 The Human Energy Budget

Since human beings are homeotherms requiring a constant core temperature near 37 °C (98.6°F), they are also required to continually manage their energy exchanges with the microclimate by physiological and behavioral adaptations and adjustments. Thus, modern studies of human thermoregulation are based on energy budgets as initially and elegantly formulated by Gagge (1936):

$$M \pm R \pm C - E = \pm S \ (\mathrm{W/m^2}), \tag{1}$$

where M is the net metabolic rate, R, C and E are radiation, convection, and evaporation respectively, while S refers to heat storage within body tissues. The flux density units in this simple energy balance equation are watts per meter squared of DuBois and DuBois (1916) body surface area.

Starting with the fundamental laws of thermodynamics, further development of detailed algorithms for the human energy balance equation represented in Eq. (1) has been a preoccupation of much biophysical and thermophysiological research over recent decades. Countless samples of subjects from different demographic and climatic backgrounds have been repeatedly tested in climate-chamber laboratories under wide ranges of ambient microclimatic conditions and with varying postures, metabolic rates, states of acclimatization, and clothing insulation. Useful reviews may be found in monographs and journal articles, including those by Hardy (1963), Fanger (1970), McIntyre (1980), Cena and Clark (1981), and ASHRAE (1993). It is sufficient here to summarize that, in the maintenance of homeothermy, the three modes of energy exchange - radiation, convection, and evaporation - can be related to just six basic and readily measured (or estimated) variables - two personal and four atmospheric:

Personal
1. Metabolic rate
2. Clothing insulation

Atmospheric
1. Air temperature
2. Thermal radiant temperature of surroundings
3. Rate of air movement
4. Atmospheric humidity

Some empirical functions describing the various modes of the dependence of heat transfer on these six basic parameters have been listed in Table 1, which can be used to fully expand the expression for skin heat loss (Q_{sk}) while still retaining its resemblance to Ohm's law. The sum of a pair of potentials, one for sensible heat and another for latent heat, is divided by their respective resistances, as seen in Eq. (2) (ASHRAE 1993):

$$Q_{sk} = \frac{t_{sk} - \dfrac{\bar{t}_r \cdot h_r + t_a \cdot h_c}{h_r + h_c}}{R_{cl} + \dfrac{1}{[(h_r + h_c) \cdot f_{cl}]}} + w \cdot \frac{p_{sk,s} - p_a}{R_{e,cl} + \dfrac{1}{(LR \cdot h_c \cdot f_{cl})}}, \tag{2}$$

where the nomenclature is as explained in Table 1. The thermal resistances forming the two denominators in Eq. (2) are each comprised of two components connected in series - one referring to the clothing layer on the skin's surface (subscript $_{cl}$), and the other representing the boundary layer of air adhering to the surface of the clothed body.

Table 1. Coefficients for various heat balance terms. (After ASHRAE 1993)

Radiation	Radiative heat flux	$R = f_{cl}\, h_r\, (t_{cl} - \bar{t}_r)$	W/m^2
	Radiative heat transfer coefficient	$h_r = \varepsilon\, 4.7$	W/(m^2 K)
	Clothed area factor	$f_{cl} = A_{cl}/A_D = 1 + 0.3\, R_{cl}$	nd
	Surface emissivity	$\varepsilon = 0.95$ unless reflective coating used	nd
Convection	Convective heat flux	$C = f_{cl}\, h_c\, (t_{cl} - t_a)$	W/m^2
	Convective heat transfer coefficient	$h_c = 8.3\, v^{0.6}$ seated ($0.2 < v < 4$ m/s)	W/(m^2 K)
		$h_c = 3.1$ seated in still air ($v < 0.2$ m/s)	W/(m^2 K)
	Respiratory convective heat flux	$C_{res} = 0.0014\, M\, (34 - t_a)\, / A_D$	W/m^2
Evaporation	Skin evaporative heat flux	$E_{sk} = w\, E_{max}$	W/m^2
	Maximum potential skin evaporative heat flux	$E_{max} = (p_{sk,s} - p_a)\, [R_{e,cl} + 1\, /(f_{cl}\, h_e)]$	W/m^2
	Evaporative heat transfer coefficient	$h_e = h_c\, (LR)$	W/(m^2 kPa)
	Lewis relation	$(LR) = 16.5$	K/kPa
	Respiratory evaporative heat flux	$E_{res} = 0.0173\, M(5.87 - p_a)/A_D$	W/m^2

$\bar{t}_r$ is mean radiant temperature of environment (°C), A_{cl} is surface area of the clothed body (m^2), M is metabolic heat generation (W/m^2), R_{cl} is clothing thermal insulation (clo), $R_{e,cl}$ is evaporative heat transfer resistance of clothes [(m^2 kPa)/W], t_{cl} is mean surface temperature of clothed body (°C), $p_{sk,s}$ is saturation vapor pressure at skin temperature (kPa), w wettedness or fraction of A_D wet with sweat (nd). $A_D = 0.202\, m^{0.425}\, h^{0.725}$ where A_D is DuBois surface area (m^2), m is body mass (kg), h is height (m) v is air movement (m/s).

2.2 Thermal Indices

The basic heat-balance equation of Eq. (1) has been further analyzed by numerous authors to produce single-unit indices of heat strain generated by conditions when $S = 0$. These indices have merit in attempting to translate departures of the various heat flux terms from their resting set-points into objective and meaningful magnitude estimates of physiological effect and subjective thermal sensation.

In cold conditions, estimation of human thermoregulatory response appears to be a relatively simple matter of linear compensation for excessive dry heat loss. For example, $S = -10$ W/m^2 can be balanced by a corresponding increase in metabolic rate, or by a corresponding diminution of radiation and convective heat losses through lowering the temperature gradient between the clothed body's surface and the ambient environment. This can easily be accomplished by interposing more thermal resistance in the form of clothing between skin temperature and ambient environmental temperature.

In warm environments where positive values of *S* exist, the latent heat flux term (E) is central to reestablishing heat balance. However, account also has to be made of the inefficiencies involved in sweating, since not all sweat produced is effective in body cooling and the excess production and subsequent dripping off the body must be accounted for by indices based on responses under heat stress. Examples of these indices include the predicted 4-h sweat rate (McArdle et al. 1947; Smith 1955), the heat-strain index (Belding and Hatch 1955), the relative strain index (Lee and Henschel 1963), and the index of thermal stress (Givoni 1963).

Szokolay (1985) has differentiated between the preceding indices and those which are empirically established using environmental psychometric methods, such as the comfort-vote technique based on sensation scales as depicted in Table 2. Empirical examples include the old effective temperature (Houghten and Yaglou 1923a, b), equivalent warmth (Bedford 1936), resultant temperature (Missenard 1935), and the equatorial comfort index (Webb 1959).

The old effective temperature (ET) and corrected effective temperature (CET) are probably the best-known empirical indices. The latest variant, known as the new effective temperature (ET*), reverts from the straight empirical approach to an analysis of heat transfers and their relationship to physiological responses such as mean skin temperature and skin wettedness (the fraction of maximum possible evaporative heat loss from a completely wet skin surface area) via the two-node thermophysiological simulator (Gagge et al. 1971; Gagge et al. 1986). ET* is currently used to define the cool and warm boundaries of the "comfort zone" within the American Society for Heating, Refrigerating and Air-Conditioning Engineers' ANSI/ASHRAE Standard 55 – Thermal Environmental Conditions for Human Occupancy (ASHRAE 1992).

Simply described, ET* represents the temperature of an isothermal environment at moderate (50%) humidity in which a lightly clothed subject would exchange the same total heat, sensible plus latent, as in the complex environ-

Table 2. Several rating scales used in thermal comfort research and empirical index formulation. The table also indicates the assumed equivalencies *(shading)* between various scales of thermal sensation, comfort preferences, and acceptability

	ASHRAE scale[a]	Bedford scale[b]	Acceptability[c]	Preference (McIntyre)[d]
3	Hot	Much too warm		
2	Warm	Too warm	Unacceptable	Want cooler
1	Slightly warm	Comfortably warm		
0	Neutral	Comfortable	Acceptable	No change
−1	Slightly cool	Comfortably cool		
−2	Cool	Too cool	Unacceptable	Want warmer
−3	Cold	Much too cool		

[a] ASHRAE (1993).
[b] Bedford (1936).
[c] Schiller et al. (1988).
[d] McIntyre (1980).

ment in question. Lines of constant ET* on the psychrometric chart are assumed also to be lines of equal thermophysiological strain in terms of mean skin temperature (t_{sk}) and wettedness (w). Due to the empirically tested correlations between t_{sk} and cold discomfort, and w and warm discomfort, loci of ET* are also assumed to be lines of constant subjective thermal sensation as well – hence their adoption as "comfort zone" boundaries in ASHRAE's Comfort Standard 55 (1992).

The widely used predicted mean vote (*PMV*) index proposed by Fanger (1967, 1970) is also a hybrid of both empirical and physically-based approaches. The vote being predicted by *PMV* is the average thermal sensation on the ASHRAE seven-point scale (Table 2) of a large sample of subjects. The equations are essentially detailed elaborations of the heat exchanges described in Eqs. (1) and (2). However, according to Fanger's theory, the achievement of homeothermy ($S = 0$) is only a necessary but not a sufficient condition for thermal comfort. In addition, Fanger contends, one needs to satisfy empirically determined comfort criteria for mean skin temperature and evaporative heat loss (a close correlate of wettedness w), which are benchmarked to the metabolic rate of the subject. Fanger demonstrated that comfortable skin temperatures have a negative linear relation to metabolic rate, while the comfortable level of evaporative heat loss has a positive linear relation to metabolic rate.

Another thermal index, derived from *PMV*, is the predicted percentage dissatisfied (*PPD*), which is an estimate of the percentage of persons in a large, random sample who would be likely to assess the thermal environment as unacceptable (assumed to equate with voting outside the central three categories of the ASHRAE seven-point scale in Table 2). The *PMV/PPD* approach indicates that acceptability for 90% of a sedentary population obtains when mean responses fall in the interval $-0.5 < PMV < +0.5$. These limits correspond to an acceptable operative temperature range of 24.5 ± 1.5 °C for persons wearing light summer clothing (0.5 clo or 0.077 $m^2\ KW^{-1}$ thermal resistance). For the same conditions, but with winter indoor clothing (1 clo or 0.155 $m^2\ KW^{-1}$ thermal resistance), the 90% acceptable range of temperatures for sedentary subjects is 22 ± 2 °C.

The *PMV/PPD* indices, together with ET* and its even more complex relative, the standard effective Temperature (*SET**), represent the most physically comprehensive attempts to quantitatively model human physiological and subjective response to thermal environment.

2.3 Localized Discomforts and Draughts

Beyond global or whole-body discomforts are the local thermal irritations such as draughts (unwanted local cooling) arising from localized air motion or proximity to a large source of radiant cooling such as a cold pane of glass. In fact, strongly asymmetric radiant environments, warm or cool, may in themselves be the source of local thermal discomfort, and can be quantified by an index known as plane radiant asymmetry (t_{pr}). Vertical temperature gradients within the occupied zone may also be irritating, and it is commonly accepted (ASHRAE 1992; ISO 1994) that a Δt of 3 °C between head and feet is as far as thermal stratification should be permitted to go before it becomes unacceptable to a significant percentage of a sedentary population indoors.

Of these local thermal discomforts, draught, or unwanted local cooling due to excessive air movement, is probably the most common source of complaint from building occupants. Not surprisingly, therefore, it has been subjected to the most intensive experimental study to date. Fanger et al. (1987) have established that the head/neck and ankle regions are the most susceptible to discomfort from draughts, and found that the percentage of a sample of sedentary subjects actually experiencing draught is related to the air temperature (t_a), mean air speed ($\bar{V}$), and turbulence intensity (TU) of the air flow. The parameter Tu can simply be defined as the ratio between standard deviation of instantaneous air speeds (V_{sd}) and the mean air speed ($\bar{V}$), both of which are derived from anemometry, having time-constants of 1/10 s or faster:

$$Tu = 100 \frac{V_{sd}}{\bar{V}} \quad x^3. \tag{3}$$

The draught risk index (*PD*), presented initially by Fanger et al. (1987) and now incorporated into ASHRAE Standard 55 (ASHRAE 1992), is written as:

$$PD = 3.143\,(34 - t_a)\,(\bar{V} - 0.05)^{0.622} + 0.3696\,(34 - t_a)\,(\bar{V} - 0.05)^{0.622}\,\bar{V}\,(Tu). \tag{4}$$

2.4 Comparison of Thermal Indices

Table 3 compares the main characteristics of selected indices. The discussion of indices thus far implies that thermal comfort is driven exclusively by physiological and biophysical processes. It would seem logical, therefore, to expect the evolution of increasingly sophisticated indices to have been accompanied by a steady increase in their correlation with comfort responses of "real" building occupants in field studies. It is perhaps surprising, therefore, that detailed field investigations have often found the opposite to be the case, at the level of an individual building occupant at least. This is shown in Table 4 by the magnitudes of correlation of common indices with thermal sensation/comfort votes (Table 2) for building occupants across climates ranging from temperate European to the equable warmth of the tropics.

This seemingly counterintuitive finding probably reflects the fact that individuals with personal preferences differing slightly from the group's mean, as one would expect within any normally distributed parameter, may in fact use clothing, metabolic rate, or air speed to bring themselves back into line with the group with whom they are forced to share a given thermal environment. That is, the interindividual differences in thermal preference are behaviorally canceled out by clothing and air speed adjustments. However, this normal variation

Table 3. Comparison of thermal indices[a]

Index	Input parameters	Applicable range	Comments
Old effective temperature (ET)	t_a, rh, v 2 levels of clothing	$0 < ET < 45$ °C $v < 2.5$ m/s	Major index, now superseded. Overestimated effects of humidity at low and moderate temperatures
Corrected effective temperature (CET)	t_a, t_r, rh, v		As above. Used in British military
Resultant temperature (T_{res})	t_a, t_{wb}, v	$20 < T_{res} < 40$ °C	Similar to ET but more accurate. Separate charts for nudesendentary and clothed-active
Equivalent temperature (T_{eq})	t_a, t_r, v	$8 < T_{eq} < 24$ $v < 0.5$ m/s	Originally measured with eupatheoscope, later calculated with Bedford's regression equation
Fanger's predicted mean vote *(PMV)* and predicted percentage dissatisfied *(PPD)*	t_a, $\bar{t}_r$, rh, v, R_{cl}, M	Moderate heat stress conditions $-2.5 < PMV < 2.5$	Less certain away from $PMV = 0$
Standard effective temperature (*SET**)	t_a, $\bar{t}_r$, rh, v, R_{cl}, M	Shivering to limit of perspiration	Based on the two-node model of thermoregulation to predict t_{sk} and w
New effective temperature (*ET**)	t_a, rh	Sedentary, light clothing	Forms the limits of summer and winter comfort zones in ASHRAE Standard 55
Draft risk *(PD)*	t_a, Tu, v	$0 < v < 0.7$ m/s	Less relevant to warm conditions above 24 °C

[a] Partly from McIntyre 1980 and ASHRAE (1992).

between individuals is ignored by the more complex indices which predict responses only for thermally average subjects. It should be remembered, therefore, that these more sophisticated indices are better suited to predictions of mean response from groups rather than individuals. Therefore, it would seem that under the ranges of conditions normally experienced by sedentary populations, and characteristically without excessive air movement, radiation, and humidity, there may actually be little reason to use other estimates of warmth beyond simple air temperature.

Table 4. Correlation of thermal comfort votes with selected indices in field investigations

Author	Sample size	Bedford seven-point comfort scale votes' correlation with. t_a	t_g	T_{eq}	*ET*	*HSI*	*ITS*	*PMV*	*SET**
Bedford (1936)	2571	0.48	0.51	0.52	0.48				
Hickish (1955)	1537	0.38	0.37	0.37	0.34				
Auliciems (1972)	2624	0.37	0.32	0.36	0.36				
Humphreys and Nicol (1970)	1284	0.67	0.66			0.62	0.59		
De Dear (1985)	3290	0.54	0.53					0.46	0.44

t_a is air temperature, t_g is globe temperature, T_{eq} is equivalent temperature, *HSI* is heat stress index, *ITS* is index of thermal stress, *PMV* is predicted mean vote, *SET** is standard effective temperature.

3 An Adaptive Model

The preceding discussion of human responses to thermal environments and the elaboration of indices of human warmth based on biophysical principles has been referred to as the "static" model of comfort (Oseland and Humphreys 1994; Nicol et al. 1995) because of its inability to deal with interindividual, intra-individual, and between-group variability in thermal response. Recently, however, discussion of an alternative model based on human adaptability and variety has appeared with increasing frequently in the literature.

3.1 Behavioral Adjustments and Preferences

Whatever the mathematics involved, homeothermy demands that, within narrow limits in Eq. (1) where $S = 0$, metabolic heat production, which varies according to activity, must be balanced by its dissipation to the environment over time. During this balancing process, all the avenues of heat transfer on the left side of Eq. (1) may be manipulated by either involuntary physiological processes controlled by the autonomic nervous system, or by voluntary behavioral responses, as listed in Table 5, which are controlled by the central nervous system. Figure 1 schematically presents the relationship between physiological and behavioral thermoregulatory systems.

As regards the direct behavioral adjustments to atmospheric ariables, there is little evidence to suggest a preference for any particular variable, despite the trend in modern air conditioning practice to exclusively manipulate air temperature whilst holding air as still as possible (<0.2 m/s), the rationale being to minimize draught risk. For example, recent evidence from detailed field experiments indicates that occupants of air-conditioned buildings in the tropics appear willing to have ambient temperature traded off against air speeds (Fig. 2), regardless of the very restricted maximum air speeds permissible under

Table 5. Physiological and behavioral thermoregulators effectors

Physiological	*Behavioral*
Vasomotor activity	Voluntary postural adjustment
Vasodilatation	Alteration of activity
Vasoconstriction	Relocation
Involuntary postural adjustment	Interposition of clothing
Cardiovascular changes	Change of diet
Thermogenesis	Migration
Muscular tension	Construction of housing
Shivering	
Sweating	
Habituation	
Acclimatization	

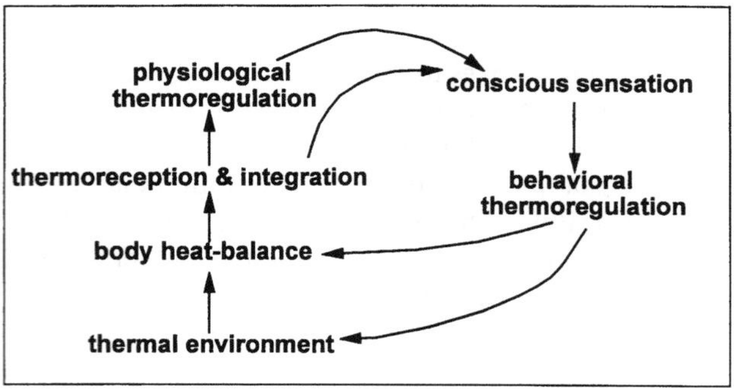

Fig. 1. Relationship between physiological and behavioral thermoregulation

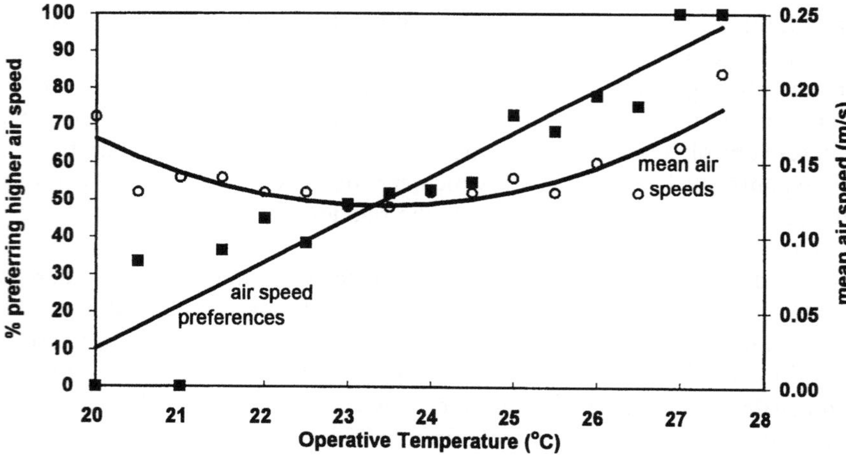

Fig. 2. Air movement preferences in air-conditioned buildings in the tropics. (After de Dear 1995)

current standards (de Dear 1995). What can be generalized at this point in time is that the trend towards ever-increasing energy inputs in the pursuit of thermal comfort in indoor climates represents a gradual substitution of engineered thermoregulation in place of behavioral and physiological means of balancing an individual's energy budget [Eq. (1)].

3.2 The Effect of Outdoor Climate

The scientific search for group differences in thermal perceptions and preferences, particularly those related to outdoor climate, has a very long history stemming from the intuitive hypothesis that persons residing in hot climates become adapted to, and actually prefer, warmer thermal environments than persons from colder climes.

Field investigations, using "real" people engaged in "real" tasks and interacting with "real" built environments, rather than highly contrived climate chamber experiments, have been regarded as the most appropriate source of evidence for adaptive effects. Field research can be classified according to the standard of instrumentation and procedure used for indoor climatic measurements. The earlier field studies measured air or globe temperature and possibly humidity, but more recent investigations have seen an evolution towards laboratory-grade instrumentation and protocols. In the latter, all of the parameters necessary for the calculation of contemporary comfort indices such als PMV, ET*, and SET have been recorded, thus enabling answers to be given to questions about how much of the variance in comfort temperatures is due to adaptive processes such as acclimatization and shifting expectations on the one hand, and, on the other, how much is simply due to the biophysics governing the body's heat balance with indoor climates. Both the earlier field studies and more recent field experiments (those with laboratory-grade instrumentation) will be reviewed below briefly.

A typical cross-sectional field study consists of administering a questionnaire with rating scales (see Table 2) to building occupants while simultaneously recording indoor climatic parameters, the most important of which is air temperature. Humidity has occasionally been included, but air velocity was usually omitted on account of the cumbersome nature of the Kata thermometer required for its measurement up until the time when hot-wire and thermistor anemometry became widely available. Numerous studies of this type have been published over the years, and Humphreys' (1975) review of 36 examples from various countries around the world uncovered a convincing statistical dependence of thermal neutralities, T_ψ, (that temperature found to correspond with "neutral" or "comfortable" thermal sensations) on mean levels of air or globe temperature (T_i) recorded within the buildings:

$$T_\psi = 2.56 + 0.83\, T_i. \quad (r = +0.96) \tag{5}$$

It was noted that building occupants were able to find comfort, assumed to be a vote on the central category of the rating scales, in indoor temperatures spanning more than 13 °C. Humphreys (1975, p. 13) attributed this to adaptive processes, concluding that "... the range of recent experience is better regarded as one of the factors which will contribute to the acceptability of the environment to which the respondent is exposed."

Since indoor temperatures are dependent on outdoor temperatures to varying extents, depending on the architecture, it seems logical that there should also be a statistical relationship between indoor thermal neutralities and outdoor climate as well. Humphreys (1976) investigated this possibility and found that the influence of external climate on indoor neutralities was particularly evident for what he termed free-running buildings, which were defined as those having neither centralized heating or colling plant. In such buildings, the following linear regression model accounted for 94% of the variance in neutralities:

$$T_\psi = 11.9 + 0.534\ T_m. \quad (r = +0.97.) \tag{6}$$

Climate-controlled buildings [those having centralized heating, ventilation, and air conditioning (HVAC) systems], on the other hand, were found to have a more moderate but still highly significant correlation with outdoor mean monthly temperature, T_m but with a curve rather than a straight line producing the best fit:

$$T_\psi = 23.9 + .295\ (T_m - 22) \cdot \exp\{-[(T_m - 22)/(24 \cdot \sqrt{2})]^2\}. \quad (r = +0.72.) \tag{7}$$

Humphreys' dataset was subsequently revised by Auliciems (1983), who deleted cases based on asymmetric rating scales, using children as subjects, and registering mean monthly temperatures below –5 °C, whilst adding more recent studies that had been overlooked or published after Humphreys' paper. These revisions brought the dataset up to 52 separate field studies from various climatic zones in Australasia, Asia, the Americas, and Europe. After combining free-running and climate-controlled buildings together, Auliciems regressed survey neutralities (T_ψ) on mean monthly outdoor temperatures T_m. The resulting linear equation was:

$$T_\psi = 0.31\ T_m + 17.6. \quad (r = +0.88.) \tag{8}$$

Incorporating an expression for prevailing indoor warmth into the model lifted the correlation coefficient to $R = 0.95$:

$$T_\psi = 0.48\ T_i + 0.14\ T_m + 9.22. \quad (R = +0.95.) \tag{9}$$

3.3 The Role of Thermal Expectation

Such empirical observations were further elaborated by Auliciems (1981) and have led to the articulation of an adaptive model, or at least a hypothesis concerning interactions between indoor-outdoor climate and perceptions of thermal comfort, as depicted schematically in Fig. 3. This model's hypothesis states that satisfaction with an indoor climate is achieved by a correct matching of actual prevailing thermal environmental conditions and one's thermal expectations of what the indoor climate should be like in that particular context. This clearly contrasts with the biophysical or static model of thermal comfort, which only looks at the actual thermal environmental conditions as they directly affect heat balance at the time of the comfort vote. As indicated in the adaptive model schematic in Fig. 3, these all-important expectations result from a confluence of one's past thermal experiences both indoors and out, along with cultural and technical practices in any given architectural and social context.

Apart from adapting physiologically and psychologically to their thermal milieu, building occupants have another option, termed adjustment in Fig. 3 and referring to direct manipulations of the biophysical parameters discussed in preceding sections of this chapter. That is, behavioral modifications of the

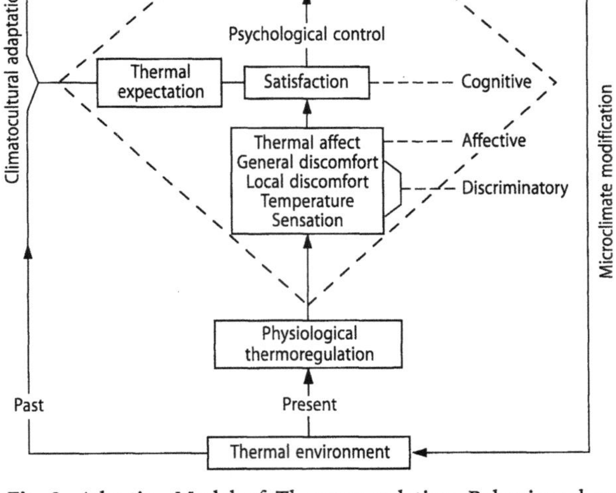

Fig. 3. Adaptive Model of Thermoregulation. Behavioural and technocultural responses to thermal discomfort are modified by acclimatization processes that determine thermal expectation. The model places empasis on psychological control that is absent from more traditional thermophysiological models (from Auliciems 1983)

body's theat balance, for example, by opening windows, turning on fans, adjusting thermostats on heating or air conditioning, and modifying clothing levels, are all termed adjustment. The behaviors have to directly impinge upon one or more of the six primary inputs to comfort models such as Fanger's PMV (ISO 1994) or two-node (Gagge et al. 1986) to qualify as thermal adjustments in this chapter. Architecture, heating, and air conditioning all represent technological and cultural adjustments to climate.

Under the adaptive hypothesis, clearly one set of indoor climatic conditions can elicit varying levels of comfort and satisfaction from building occupants, depending on the climatic, technological, and cultural expectations of the latter. By logical extension, the hypothesis also implies that the temperatures people expect indoors for comfort and satisfaction will drift in the direction of the average conditions encountered in their day-to-day life, both indoors and out, since the two realms interact trough climatically permeable architecture.

So, in the language of cybernetics, outdoor climate acts as negative feedback which attracts the thermal perceptual system's set point, thereby damping load error (dissatisfaction or discomfort) within the behavioral thermoregulatory feedback loop. In the parlance of the impact models (see Kates 1985), the adaptive hypothesis argues that involuntary behavioral responses are not restricted to second-order impacts, but also refer to acceptance of some level of first-order impacts, albeit as modified by the thermal expectation feedback from third-order impacts.

3.4 Testing the Adaptive Model

While the relationship between neutralities and prevailing outdoor climate appears convincing according to empirical correlation coefficients, the actual causal mechanism is left in doubt by the discussion in preceding paragraphs. Apart from adaptive processes of shifting expectations (call it habituation) and acclimatization, there are several alternative, simpler hypotheses, including the possibility that some parameters in the human body's heat balance which were not measured in the field studies were in fact compensating for indoor temperatures which were measured. That is, the dependence of thermal neutralities on outdoor climates was simply the result of biophysical adjustments and not adaptation at all. For example, it may be possible that lower clothing, metabolic, and humidity levels may combine with higher air velocities in the warmer climates to allow subjects to attain thermal comfort and neutrality in considerably higher indoor temperatures than would otherwise have been feasible. Therefore, to more rigorously test the adaptive hypothesis, these simpler alternatives need to be addressed. To that end, a type of field investigation has evolved over the last decade in which the full complement of biophysical measurements are taken with instrumentation approaching laboratory-grade. As noted above, such investigations can more accurately be described as "field experiments" (de Dear et al. 1991).

De Dear and Auliciems (1985a) reported on thermal comfort experiments in various Australian climate zones ranging from equatorial (Darwin) through subtropical (Brisbane) to midlatitude (Melbourne). In both Melbourne and Brisbane, two experiments were conducted during their respective summer seasons, one in free-running buildings and the other in climate-controlled buildings. In total, these Australian samples included over 1100 office building occupants who cast thermal assessments and had their indoor climatic environments measured on 3290 separate occasions. The six field experiments' neutralities, estimated by the statistical technique of probits, revealed an obvious dependence on mean monthly outdoor temperatures, but the trend was most pronounced in the free-running buildings in Brisbane and Melbourne. In fact, the Brisbane free-running building sample had the warmest neutrality at 25.6 °C, while Melbourne's free-running sample had the coolest at 21.8 °C. It is unfortunate that the investigators were unable to locate a sample of free-running office buildings in Darwin, where the effect of a hot and humid tropical climate could be expected to be most evident. The air-conditioned buildings across Australia on the other hand all had neutralities clustered within the relatively narrow range between 23 and 24 °C.

Schiller et al. (1988, 1990) tested the ASHRAE Standard 55 in a pair of field experiments during the summer and winter seasons of the San Francisco Bay Area. The instruments used in their investigations can be regarded as laboratory-grade, with simultaneous readings from three heights (0.1, 0.6, and 1.1 m) above floor by air and globe thermistors, omnidirectional anemometers plus plane radiant asymmetry and dew-point sensors. Concurrent subjective responses from the building occupants were logged directly onto a questionnaire administered to the subjects on a laptop computer.

De Dear and Fountain (1994) extended the Schiller et al. (1988, 1990) study to a hot and humid climate, with two detailed field experiments using similar laboratory-grade instrumentation and identical protocols. Their samples were taken during the wet and dry seasons in Townsville in Australia's tropical north. For comparison with other field experiments, de Dear and Fountain (1994) derived neutralities from the seven-point ASHRAE thermal sensations scale (Table 2) applied within air-conditioned buildings and found them to be between 24 and 25 °C. Interestingly, it should be noted that this was a full degree warmer than the temperatures the subjects found most acceptable or preferable, suggesting that the long-assumed equivalence between the various thermal scales depicted in Table 2 may, in fact, be unjustified.

Busch's (1990) experiments in various office buildings in tropical Bangkok used much the same basic methods as de Dear and Auliciems' Australian experiments (1985), with air and globe temperatures, humidities, and air velocities being registered at a sample of workstations simultaneously with the completion of thermal questionnaires by the occupants of those workstations. Both climate-controlled (air-conditioned) and free-running (naturally ventilated) buildings were studied, so a diversity of architectural contexts was included in the final sample size of 1146. Separate probit analyses were conducted for the climate-controlled and free-running samples, and the neutrality in the former was found to be 24.5 °C, which was quite consistant with the values obtained by de Dear and Auliciems in air-conditioned buildings in the Australian tropics and stubtropics (1985a). However, in Bangkok's free-running buildings, Busch observed a neutrality of 28.5 °C, a result which was over three degrees warmer than predicted by Fanger's PMV index based on purely biophysical factors alone, and assuming Busch's clo estimate plus 0.15 clo chair insulation (McCullough and Olesen 1994) for the office workers seated at their desks at the time of the questionnaire. Auliciems' (1983) adaptive model [Eq. (9)], on the other hand, came within half a degree of the Bangkok observation.

De Dear et al. (1991) conducted experiments in air-conditioned office buildings and free-running residential apartment blocks in equatorial Singapore. Instrumentation consisted of Assmann aspirated psychrometers, hot-wire anemometers and globe thermometers, along with a pencil-and-paper questionnaire/interview which was administered by a local research assistant fluent in all the major languages of Singapore. The observed neutrality of 24.2 °C in the air-conditioned buildings was, again, quite consistent with the Darwin and Bangkok results reported above, and accurately predicted by both the adaptive and heat-balance models once the 0.15 clo chair insulation was added to the original authors' clothing estimates (de Dear 1994). As was the case in Busch's Bangkok experiment, the 28.5 °C neutrality observed in Singapore's free-running buildings was underestimated by both models, but again the adaptive model came closest with a predicted neutrality of 27.2 °C.

For comparative purposes and an examination of the climate factor's potential influence on thermal perception indoors, a selection of these comfort field experiments has been compiled into a graph showing observed neutralities in relation to mean monthly outdoor temperature (Fig. 4). Also plotted on the same graph are the neutralities predicted on the basis of purely heat-balance factors. The PMV model is an appropriate index here since it enables the effects of not only air temperature, but also all the other relvant biophysical factors discussed above to be taken into account. An important point to note about these PMV calculations is that they have taken into account the incremental insulation of chairs, even if the original authors had omitted this effect. Superimposed on the heat-balance predicted neutralities is a set of predictions from the adaptive model which has mean indoor and outdoor monthly temperatures as the inputs [Eq. (9)].

It would appear as though, in these tropical field experiments at least, the adaptive and the heat-balance models performed equally well when predicting thermal neutrality in the air-conditioned buildings. However, the elevated neutralities observed in the naturally ventilated building experiments in Bangkok and Singapore were significantly better predicted by the adaptive model of Eq. (9).

This raises the question of why outdoor climate influences occupants' thermal expectations of free-running buildings but not air-conditioned buildings? If we accept that it is one's total thermal history that determines expectation, as

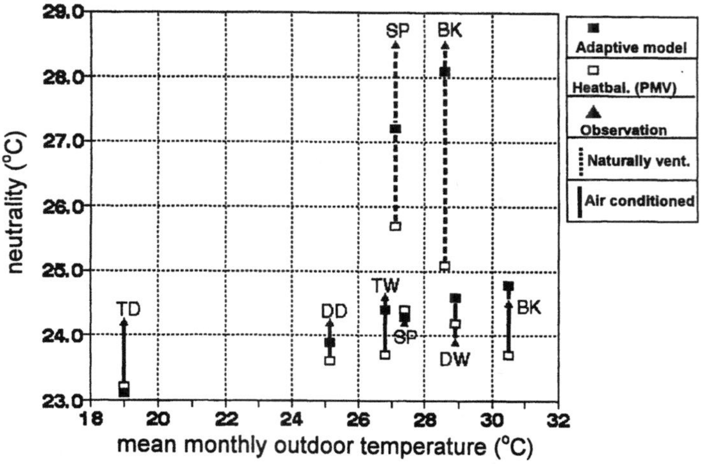

Fig. 4. Adaptive and heat-balance model predictions compared with neutralities observed in eight tropical field experiments. *DD* and *DW* Darwin dry and Darwin wet respectively (de Dear and Auliciems 1985a), *SP* Singapore (de Dear et al. 1991), *BK* Bangkok (Busch 1990), *TD* Townsville dry, *TW* Townsville wet (de Dear and Fountain 1994)

implied by the schematic in Fig. 3, then exposure to air-conditioned indoor climates presumably biases expectations towards the air-conditioning set points, which under today's indoor climate standards of thight control remain locked in to the 22–24 °C range due to their derivation from heat-balance comfort models such as PMV in which all the input parameters have been standardized. On the other hand, the indoor climatic expectations engendered by free-running buildings must be more closely correlated with outdoor temperatures – people recognize that their building is naturally ventilated and relax their expectations accordingly, becoming more tolerant of indoor temperatures that are dynamically linked with diurnal, synoptic, and seasonal cycles in outdoor climate. Rohles and Wells (1977) alluded to this same point in his explanation of why Michigan experimental subjects were more tolerant of 32° ET* in a climate chamber exposure than Texans – it was speculated that the latter took summer air conditioning for granted, and that led them to be much more critical of elevated warmth indoors than their northern counterparts who had not yet had their thermal expectations elevated to the same stringent levels as the Texans.

In their recent tropical building field experiment, de Dear and Fountain (1994) found Townsville office occupants to be so finely "tuned" to the thermostatic and isothermal indoor climates of their air-conditioned buildings that their thermal sensations were observed to be at least twice as sensitive to temperature than expected on the basis of previous experiments. As depicted in Fig. 5, the gradient of the actual mean thermal sensation (n = 1234) in relation to indoor operative temperature was significantly steeper than predicted by a biophysical heat-blanace index (PMV) which accounted for not only tempera-

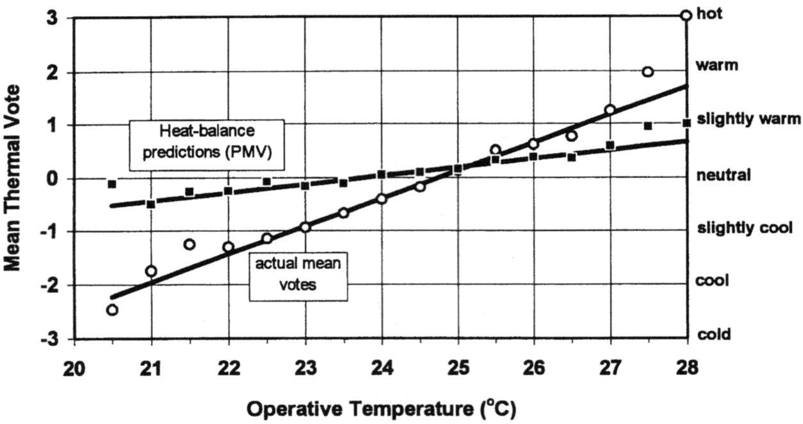

Fig. 5. Heightened sensitivity of thermal comfort ratings in tightly controlled air-conditioned buildings in the tropics (after de Dear and Fountain 1994). The gradient of the actual mean votes (within 0.5 °C) on indoor operative temperature was 0.5 vote units per °C, which was significantly steeper than that predicted by a biophysical comfort model (PMV) which took into account clothing insulation, metabolic rates, humidity, and air speed as well as operative temperature

tures but also air speed, humidity, metabolic rate, and clothing insulation. The implication of this finding is that the occupants of these tightly controlled air-conditioned buildings had become hypersensitive to departures from thermal expectation, and although they were behaviorally adapting to such departures by modifying their clothing insulation (as implied by the shallow gradient on the PMV regression), one can speculate that they felt that it was not their job to do any thermoregulation whatsoever. As one subject succinctly put it: "... After all, that's what air-conditioning is for, isn't it?"

Similar anecdotes abound in Britain as well, where the practice of summer airconditioning in offices is a relatively recent phenomenon. British architects and engineers now find themselves on an "indoor climate treadmill" with ever-increasing expectations from building owners and managers who have come to regard air conditioning as the norm in that particular market nowadays, despite the relatively trivial number of cooling degree-days experienced in the British climate. However, it would be premature to conclude from such anecdotes and speculations that indoor climate expectations can only go in one direction – up. There must be also be scope for expectations to go down, or rather, become relaxed, as well, since expectations are cognitive constructs and as such, amenable to deliberate manipulation.

4 An Evaluation of Air Conditioning

The preceding sections raise important issues regarding human comfort and possibly physiological fitness and health, particularly as they relate to the need to reduce external energy consumption within the built environment. Not the least is the concern for unnecessary emission of radiatively active gases into the global atmosphere. The central task, therefore, is to establish the most energy-efficient management strategy for indoor comfort control which will be simultaneously compatible with the well-being and expectations of the building occupants. These general concerns with tightly controlled air conditioning are illustrated by the flow chart in Fig. 6.

To summarize, firstly, the recirculation of air in air-conditioned systems is not particularly efficient at removing certain noxious substances. Indoor pollutants, as further discussed by de Dear (this Vol.), may accumulate to unhealthy concentrations. Moreover, several components and processes in modern HVAC

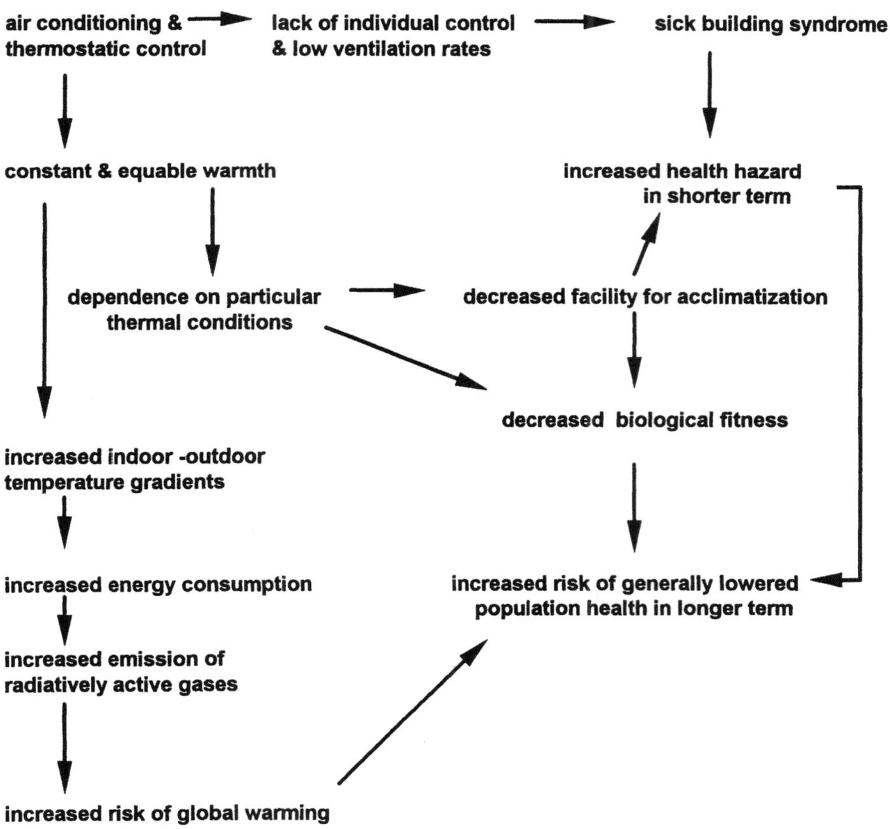

Fig. 6. Negative consequences of excessive reliance on thermostatically controlled air conditioning

systems increase the potential for growth of harmful microorganisms. Secondly, air conditioning, closely, or thermostatically, controlled, tends to produce dependable flows of isothermal air. Both the adaptive theory and practice tend to suggest that people may actually become dependent upon such persistently equable conditions. This, in turn, further reinforces expectations, leads to even closer attention being paid to the efficiency of such operations. Both in preference and clothing, standardization is enforced, and the capacity for individual adjustment and adaptation is further diminished. Sustained over the longer period, such reliance is likely to reduce the occupant's dynamic range of physiological and behavioral thermoregulatory response.

An unfortunate extension of this process of habituation produces an increased positive feedback that demands ever larger spaces to be air-conditioned: sports arenas, shopping complexes, both public and private modes of transport. The sum total of human contact with the continuous variability of the natural environmental is ultimately minimized. In a highly entertaining essay expressing concern for North Americans' addiction to air-conditioned air, Prins (1992) described the situation as "... the most pervasive and least noticed epidemics in modern America" and coined the label "Condis" to refer to its victims. A "Condi" is cold in summer and hot in winter; Prins may well have been too optimistic in assuming that they were confined just to the North American continent.

In the "big picture", whatever the arguments in favor of such developments, the uncriticial proliferation of static air conditioning holds also the potential for a progressive degeneration of thermoregulatory capacity in people living in tightly controlled environments. Should sudden disruptions occur in their provision and maintenance, as may well be the case after significant global warming has occurred, hypersensitized and habituated populations are likely to be exposed to degrees of thermal stress which they are ill-prepared for physiologically.

Traditionally, coping with thermal inclemency has involved modifications in lifestyles to suit particular diurnal, synoptic, and seasonal climatic rhythms. For example, adoption of siestas during the hottest part of the day represents a climatically appropriate lifestyle adaptation. Indoor thermal regulation has depended more on passive energy technologies such as natural ventilation, shading, sizing and positioning of openings, solar orientation of buildings on their sites, selection of appropriate materials, and deliberately exploiting the thermal mass of the building envelope. Moreover, at this time there also exists more than just prima facie evidence to suggest that, at least in some instances, such traditional and passive strategies may actually be preferable to individuals in today's energy- and greenhouse-conscious world. At least the rationale and economy of air cooling in latitudes above 35° needs to be put under closer scrutiny. The moral dimension to this discussion becomes even starker when one remembers that the wealthy countries of those same latitudes are currently encouraging the countries of the less-developed world to curb their appetites for energy and its attendant greenhouse gas emissions. What is questioned here, however, is not air conditioning per se, but rather its indiscriminate application in areas and at times when other technologies are available, economically viable, and habitable.

4.1 Achieving Acceptance of Adaptive Control Strategies

From the preceding theoretical discussions, general considerations of air-conditioning management practices, and end-user surveys, there would seem to exist an overwhelming case for reconsidering indoor climatic control strategies. In general, the approach needs to gradually "wean" people who are currently "addicted" to equable warmth (Prins, 1992), and to ensure that this is achieved economically and without diminution of overall thermal satisfaction levels.

One approach could be as follows: for populations already residing in buildings with closely controlled indoor climates, a program of deregulation could be instituted by allowing a gradual drift of indoor temperature set-points towards outdoor temperatures, as described in Eq. (6–9). This is likely to encourage greater adjustment of individuals' clothing insulation and the wearing of ensembles that better reflect weather conditions outdoors. Building occupants in warmer climatic zones, where the scope for thermal insulation adjustment is restricted by culturally determined dress-codes, would be expected to exploit forced-convective heat fluxes in their heat-balance equation [Eq. (1)] by increasing air speeds above the "still air" specifications enforced by current air-conditioning standards (ASHRAE 1992). Both the clothing and air-speed responses to dynamic indoor climatic conditions, in turn, are likely to further reduce the thermal "shock" of moving from one environment to another, and presumably promote more frequent outdoor activities by building occupants. Therefore the prospects for seasonal acclimatization would also be enhanced. The overall result should be a closer matching of indoor thermal expectations to levels of warmth prevailing outdoors.

The problem remains a relatively simple one: how best to continuously adjust set-points for heating and air-conditioning systems?

4.2 Adaptive Alternatives to the Thermostat

During the past decade, microprocessor developments have enabled the already versatile thermostat to undertake a remarkably sophisticated control role. In an effort to conserve energy and decrease the adjustment time of conditions indoors, the thermostat has been variously modified to anticipate transients in outdoor and indoor heat loads, to setback control targets, manipulate control deadbands at predetermined times, and in general to achieve an increased precision in operations. By definition, however, and irrespective of the complexity of control logic or designation of purpose, any thermostat can aim only to achieve a stasis within the internal environment. Its gyrations per se are no more than a means of achieving temperature constancy according to some predetermined comfort criteria which, itself, has been usually regarded as static. However, the thermostat function can be readily altered to a dynamic one simply by programming additional climatic data inputs and integration by algorithms such as the adaptive equations discussed earlier (Auliciems 1981; Humphreys and Nicol 1995). The resulting control systems might be referred to as thermoadapters, or, more simply as thermobiles (Auliciems 1981). The only major uncertainty with this proposal, however, concerns the rate of "weaning" from thermostatic control towards fully adaptive strategies. There already exists ample field evidence indicating that a "sudden" switch-over would not be well received by building occupants (Stolwijk 1978), since thermal expectations would still remain firmly rooted in the static thermal regime. Clearly, further field experimentation with real buildings and their occupants is required, with the focus being on the quantification of time constants for "thermal expectations" and the establishment of the most effective means for their modification prior to full implementation of adaptive thermal control strategies. Sensibly, however, occupants of already existing buildings should be fully informed of the purpose of the changeover to adaptive control and their full support enlisted from the outset of any such program. Likewise, with new buildings in which adaptive control strategies are to be practiced, the occupants need to be fully advised of what to expect from the building's indoor climate and thoroughly educated in their behavioral thermoregulatory possibilities. In short, occupants of adaptive buildings need to be taught how to once again assume individual responsibility for thermoregulation and not to expect air-conditioning systems to perform 100% of the job for them, as is the case in conventional thermostatically controlled buildings.

Thermobiles with their adaptive control logic could be incorporated into all new building designs where active energy usage for occupant thermal comfort is required, and indeed, could readily replace existing thermostats. Alternatively, the existing thermostatic devices may well be amenable to conversion to achieve the desired adaptive control. In any case, the cost of this modification should be minimal. The most fundamental modification necessary for implementation of adaptive control is the incorporation of outdoor temperature data into the control circuitry. Outdoor temperature sensors are already in place with many modern air-conditioning systems which have economizer cycles built in – i.e., systems which increase the intake of outdoor air whenever weather conditions are sensed within a predetermined range.

The resulting benefits of adaptive controls would be twofold. First, overall comfort and thermoregulatory fitness levels of building occupants should be enhanced. Second, and perhaps even more importantly, the reductions of indoor-outdoor thermal gradients would lead to savings in energy. For example, the adherence to standardized dress codes in thermostatically controlled air-conditioned office buildings in the United States was recently estimated by the Rocky Mountains Institute to cost the average business US$ 1860 per annum in air-conditioning energy above what would be the case if occupants were encouraged to adapt clothing patterns to synoptic and seasonal climatic cycles (Chronicle 1995). Obviously, on a worldwide basis, the potential for energy conservation and reductions in greenhouse gas emission offered by adaptive control is enormous.

5 Summary

1. People cope with moderate thermal discomfort and stress with a combination of physiological and behavioral responses. These responses are dynamic and occur at time scales ranging from immediate adjustments through to longer-term adaptations (habituation) within an individual's life span and beyond (culture).
2. Thermal comfort is a complex psychophysiological state of satisfaction with indoor climate that occurs across a narrow range of these responses.
3. Levels of thermal comfort achieved indoors may be predicted for a given population from metabolic rate, clothing, ambient indoor thermal parameters, and levels of outdoor warmth as varying with weather and season.
4. Optimization of thermal comfort depends on the conditions of testing: extrapolations of results from laboratory studies to real-life situations may not be valid. Therefore, specification of optimal indoor levels of warmth requires validation in the field.
5. In general, however, comfort levels will vary with prevailing levels of warmth both indoors and out. Therefore, prescriptions for optimal indoor conditions need to rely on the expressed thermal preference for any population, depending on its outdoor climatic context, prevailing technological practices, and thermal expectations.
6. From a climatic viewpoint, air conditioning at the present time is largely overused, both with respect to geographic location and prevailing levels of seasonal warmth.
7. General human responses to air conditioning are inadequately understood at present, and little is known about cultural or regional differences in preferences. Clearly the current focus on temperature control to the exclusion of other significant parameters in the overall thermal environment seems to be a gross oversimplification of indoor climatic management.
8. Overall levels of satisfaction with the performance of contemporary air-conditioning systems within both homes and office buildings is low.
9. Air conditioning management strategies are overly reliant on simple "brute-force" energy-intensive solutions, and often incompatible with global environmental responsibilities and end-user requirements.
10. New approaches to the management of indoor climate should explore alternatives to thermostatic control. In particular, attention should be paid to more harmonious synchronization of indoor and outdoor levels of warmth and establishment of controlled variability of indoor atmospheric conditions to foster increased human adaptation.

References

American Society of Heating, Refrigerating, and Air Conditioning Engineers (ASHRAE) (1992) Standard 55-92: Thermal Environmental Conditions for Human Occupancy. ASHRAE, Atlanta, Georgia
ASHRAE (1993) Handbook of Fundamentals. ASHRAE, Atlanta, Georgia
Auliciems A (1981) Towards a psycho-physiological model of thermal perception. Int J Biometeorol 25:109–122
Auliciems A (1983) Psycho-physiological criteria for global thermal zones of building design. Int J Biometeorol 26 (Suppl):69–86
Bedford T (1936) The warmth factor in comfort at work. Report – Industrial Health Research Board 76. HMSO London
Belding HS, Hatch TF (1955) Index for evaluating heat stress in terms of resulting physiological strain. Heat Pip Air Condit 27:129–136
Benzinger TH (1978) In introducing part III of the First International Indoor Climate Symposium, Aug 30 to Sep 1, Copenhagen
Burton AC, Edholm OG (1955). Man in a cold environment. Edward Arnold, London
Busch JF (1990) Thermal responses to the Thai office environment. ASHRAE Trans 96:859–872
Cena K, Clark JA (1981) Bioengineering. Thermal physiology and comfort. Elsevier, Amsterdam
Chappell JE (1975) The ecological dimension: Russian and American views. Ann Assoc Am Geogr 65:144–162
Chronicle (1995) Casual dress days. San Francisco Chronicle, 21 August, D4, Chronicle Publishing Co. San Francisco
Collins KJ (1979)Hypothermia and thermal responsiveness in the elderly. In: Fanger, P, Valbjørn O (eds) Indoor Climate: Proceedings of the First International Indoor Climate Symposium. Danish Building Research Establishment, Copenhagen
De Dear RJ (1985) Perceptual and adaptational bases for the management of indoor climate. Ph D Thesis, University of Queensland, Brisbane
De Dear RJ (1994) Outdoor climatic influences on thermal requirements indoors. In: Oseland NA, Humphreys MA (eds) Thermal comfort: past, present and future. Building Research Establishment, Garston UK, pp 106–132
De Dear RJ (1995) Thermal comfort in air-conditioned office buildings in the tropics. In: Nicol F, Humphreys M, Sykes O, Roaf S (eds) Standards for thermal comfort: indoor air temperature standards for the 21st century. E & FN Spon – Chapman and Hall, London, pp 122–131
De Dear RJ, Fountain ME (1984) Field experiments on occupant comfort and office thermal environments in a hot-humid climate. ASHRAE Trans 100:457–475
De Dear RJ, Auliciems A (1985a) Validation of the predicted mean vote model of thermal comfort in six Australian field studies. ASHRAE Trans 91:452–468
De Dear RJ, Auliciems A (1985b) Thermal neutrality and acceptability in six Australian field studies. In: Fanger PO (ed) Clima 2000, 4. VVS Kongres-VVS Messe, Copenhagen, pp 103–108
De Dear RJ, Leow KG, Foo SC (1991) Thermal comfort in the humid tropics: field experiments in air conditioned and naturally ventilated buildings in Singapore. Int J Biometeorol 34:259–265
DuBois D, DuBois EF (1916) A formula to estimate approximate surface area, if weight and height are known. Arch Intern Med 17:863–871
Dubos R (1980) Man adapting. 2nd ed. Yale University Press, New Haven
Fanger PO (1967) Calculation of thermal comfort: introduction of a basic comfort equation. ASHRAE Trans 73:111.4.1–111.4.20
Fanger PO (1970) Thermal comfort. Danish Technical Press, Copenhagen
Fanger PO, Melikov A, Hanzawa H, Ring J (1987) Air turbulence and sensation of draft. Energy Buildings 12:21–39
Fox RH, Woodward PM, Exon-Smith AN, Green MF, Donjnison, DV, Wicks MH (1973) Body temperature in the elderly: a national study of physiological, social and environmental conditions. Br Med J 1:200–206

Gagge AP (1936) The linearity criterion as applied to partitional calorimetry. Am J Physiol 116:656–668
Gagge AP, Stolwijk JAJ, Nishi Y (1971) An effective temperature scale based on a simple model of human physiological regulatory response. ASHRAE Trans 92:709–731
Gagge AP, Fobelets A, Berglund LG (1986) A standard predictive index of human response to the thermal environment. ASHRAE Trans 75:108–125
Givoni B (1963) Man, climate and architecture. Elsevier, Amsterdam
Hardy JD (ed) (1963) Temperature, its measurement and control in science and industry, vol 3. Reinhold, New York
Hickish DE (1955) Thermal sensations of workers in light industry in summer: a field study in southern England. J Hyg (Lond) 53:112–123
Houghton FC, Yaglou CP (1923a) Determining lines of equal comfort. ASHVE Trans 29:163–175
Houghton FC, Yaglou CP (1923b). Determination of the comfort zone. ASHVE Trans 29: 361–379
Humphreys MA (1975) Comfortable indoor temperatures related to the outdoor air temperature. Building Research Establishment (note PD 117/76), UK Dept. of Environment, London
Humphreys MA (1976) Comfortable indoor temperatures related to outdoor air temperature. Building Research Establishment (note PD 117/76), UK Dept. of Environment, London
Humphreys MA, Nicol JF (1970) An investigation into the thermal comfort of office workers. J Instn Heat Vent Eng 38:181–189
Humphreys M, Nicol F (1995) An adaptive guideline for UK office temperatures. In: Nicol F, Humphreys M, Sykes O, Roaf S (eds) Standards for thermal comfort. E&FN SPON, London, pp 190–195
Huntington E (1924) Civilization and climate. Yale University Press, New Haven
Huntington E (1926) The pulse of progress. Simpson, New York
International Standard (1994) ISO 7730: Moderate thermal environments. International Standards Organization, Switzerland
Jones WHS (trans) (1923) Airs, Waters and places by Hippocrates of Cos. Heinemann, London
Kates RW (1985) The interaction between climate and society. In: Kates, RW, Ausubel J, Berberian M (eds) Climate impact assessment: studies of interaction between climate and society. Wiley, New York, pp 3–36
Lamb HH (1982) Climate history and the modern world. Methuen, London
LBL (1995) Lawrence Berkeley Laboratories – Indoor Environment Program. (http://eande.lbl.gov/CBS/IEP.html).
Lee, DHK, Henschel A (1963) Evaluation of thermal environment in shelters. Division of Occupational Health Technical Report 8, US Dept of Health Education, and Welfare, Cincinnati
Markham SE (1947) Climate and the energy of nations. Oxford University Press, London
McArdle B, Dunham W, Holling HE, Ladel W, Scott JW, Thompson ML, Weiner JS (1947) The predication of physiological effect of warm and hot environments. Medical Research Council Report 47-391. HMSO, London
McCullough EA, Olesen BW (1994) Thermal insulation provided by chairs. ASHRAE Trans 100:795–802
McIntyre, DA (1980) Indoor climate. Applied Science Publishers, London
Mills CA (1946) Climate makes the man. Gollan, London
Missenard A (1935) Theorie simplifié du thermometre résultant. Chauf Vent 12:347–352
Missenard A (1957) In search of man. Hawthorn Books, New York
Monteith JL, Mount LE (1974) Heat loss from animals and man: assessment and control. Butterworth, London
Newman MT (1953) The application of ecological rules to racial anthropology of the aboriginal world. Am Anthropol 55:311–327
Newman MT (1955) Adaptation of man to cold environments. Evolution 9:101–105
Nicol F, Humphreys M, Sykes O, Roaf S (eds) (1995) Standards for thermal comfort – indoor air temperature standards for the 21st century. Chapman and Hall, London
Nishi Y, Gagge AP (1977) Effective temperature scale useful for hypo- and hyperbaric environments. Aviat Space Environ Med 48:97–107

Oseland NA, Humphreys MA (eds) (1994) Thermal comfort: past, present and future. Building Research Establishment, Garston, UK
Prins G (1992) On Condis and coolth. Energy Buildings 18:251–258
Rohles FH, Wells MV (1977) The role of environmental antecedents on subsequent thermal comfort. ASHRAE Trans 83:21–29
Sargent F II (1963) Tropical neurasthenia: giant or windmill? In: Atid zone research. Environmental psychology and physiology: Proceedings of the Lucknow Symposium. UNESCO, Paris
Sargent F II, Tromp SW (eds) (1964) A survey of human biometeorology. WMO technical note no 65, Geneva
Schiller GE (1990) A comparison of measured and predicted vomfort in office buildings. ASHRAE Trans 96:609–622
Schiller GE, Arens E, Bauman F, Benton C, Fountain M, Doherty T (1988) A field study of thermal environments and comfort in office buildings. ASHRAE Trans 94:280–308
Scholander PR (1955) Evolution of climatic adaptation in homeotherms. Evolution 9:15–26
Scholander PF (1956) Climatic rules. Evolution 10:339–340
Smith FE (1955) Indices of heat stress. Medical Research Council Memo 29. HMSO, London
Stolwijk JAJ (ed) (1978) Energy conservation strategies in buildings – comfort, acceptability and health. JB Pierce Foundation, New Haven
Szokolay SV (1985) Thermal comfort and passive design. In: Boer KW (ed) Advances in solar energy 2. American Solar Energy Society, New York
Taylor GT (1959) Australia: a study of warm environments and their effect on British settlement. 7th edn. Methuen, London
Toynbee AJ (1945) A study of history, vol. 2 3rd edn. Oxford Unversity Press, London
Webb, CG (1959) An analysis of some observations of thermal comfort in an equatorial climate. Br J Ind Med 16:297–310

Sick Building Syndrome and Appropriate Design

R. de Dear

1 Introduction

An increasing number of complaints about discomfort and adverse health effects in connection with time spent in non-industrial buildings has been reported in the last 20 years. Typically the symptoms intensify during the working day within the building and then deminish or disappear entirely after leaving the building in question. This chapter reviews the literature on the subject of sick building syndrome (SBS), highlighting the various causal hypotheses that have been proposed over the two decades since the term "sick building syndrome" was coined, and suggesting possible design strategies to mitigate the problem.

1.1 Definitions

Anecdotal reports of problem buildings have appeared with increasing frequency since the early 1970s. The term "sick building syndrome" is now widely used to describe a non-specific array of symptoms which can be grouped under four broad categories; mucous membrane irritation, neurotoxic disorders, skin irritation and odour/tase complaints (WHO 1983; Molhave 1987; Hodgson 1992):

Mucous membrane irritation (eye, nose and throat):
- Irritation of eyes
- Nasal congestion
- Dry throat and cough

Neurotoxic disorders:
- Headaches
- Dizziness
- Fatigue
- Nausea
- Irritability
- Lack of concentration
- Malaise

Climatic Impacts Centre, Macquarie University, Sydney, NSW 2109, Australia

Skin irritation:
- Skin rash
- Itchiness

Odour:
- Odour and taste complaints

Despite being something of a misnomer (the occupants rather than the buildings are supposedly sick) and the existence of various synonyms such as "tight building syndrome", the term "sick building syndrome" has been accorded official World Health Organisation recognition since 1983 (WHO 1983). While it is possible for domestic residences to be associated with SBS-type symptoms (e.g. Ryd 1991), the focus of this chapter, along with the vast majority of SBS literature before it, is on the office environment.

The criterion linking such a diverse battery of complaints to office buildings is simply that the sufferers report an improvement when they leave their workplace. When applied in a research context, this building-relatedness part of the SBS definition raises risks of bias. Respondents in many published SBS surveys may well have found questionnaire items asking if symptoms were alleviated upon leaving their office building to be quite leading. Furthermore, such questionnaires often fail to distinguish between physical and social aspects of the work environment when they ask respondents if symptoms improved upon leaving the workplace, so occupational stress may be easily be mistaken for building-related SBS.

As to the question of how many sufferers a building needs before the term SBS can be applied, two decades of research have brought us no closer to a clear distinction between "sick" and "healthy" buildings, and the background level of SBS symptoms prevalence remains largely unknown. The World Health Organisation simply defines a sick building as one in which there is a "significant excess" frequency of complaints among occupants (WHO 1983) compared to the general community level; a working threshold value cropping up frequently in the literature is 20%, although whether the 20% excess is comprised of individual symptoms or some aggegate index varies widely between SBS researchers. Furthermore, it appears that virtually every office building has at least some occupants suffering from symptoms on the SBS checklist, and a higher prevalence of symptoms within a building does not necessarily mean the building is "sick". For example, of the ten Sydney office buildings surveyed by Rowe and Wilke (1994), the one described as a sick building by its occupants prior to the investigation ended up scoring close to the entire sample's mean level of SBS symptom prevalence.

A term related to SBS is building-related illness (BRI). What distinguishes BRI from SBS is the presence of a specific infectious agent or allergen in the indoor environment which can be clinically diagnosed. This leaves the SBS with something of a negative definition in that it can only be diagnosed once all other specific BRIs with aetiology have been eliminated (Molina et al. 1989). Legionnaire's disease and hypersensitive pneumonitis are the most frequently cited BRI examples, but asthma may also fall into the BRI category by virtue of the role played by allergens like the house dust mite in triggering attacks. This chapter will not deal with BRI.

1.2 Significance

The significance of the SBS problem lies not so much in the gravity of the symptoms themselves, for at worst they rate as "low grade" and are unlikely to cause lasting physical damage (Anonymous 1991; Raw 1992), but rather in the sheer size of the workforce at risk, and the attendant productivity decrements, absenteeism and personnel turnover costs. Given that over half of all workers in the economically developed world work in offices, the potential economic impacts of SBS are anything but trivial.

As mentioned earlier, an SBS-type symptom incidence of 20% or higher among building occupants is a common diagnostic criterion for "sick building", in recognition that there are complaints of this type in virtually all buildings. Assuming the prevalence of SBS symptoms is normally distributed within the population of office buildings, those buildings with an excess of complaints may be regarded as simply the tail end of the bell-shaped distribution. What remains unknown is just what percentage of office buildings fall into this tail. A carefully conducted study of the background level of SBS in the population of office buildings in any country is yet to be conducted, but some papers have suggested that as many as 30–50% of all new and refurbished office buildings with mechanical ventilation or HVAC (heating, ventilating and air conditioning) systems have "high" rates of SBS-type complaints (Wilson and Hedge 1987; Woods 1989).

Extrapolating from such numbers to aggregate economic impacts of SBS, as mediated via productivity decrements, personnel turnover and absenteeism, is dubious to say the least. However, there is something to be gained from attempting to assign "ball park figures" to the SBS economic penalty, if for no other reason than to underscore the importance of directing further research resources at this problem. Woods (1989) suggested that at any one time 10–25 million workers in 800 000 to 1.2 million commercial buildings in the United States will have symptoms typical of the SBS. Based on estimates such as these, Axelrad (1989) suggested that the economic costs from medical visits, sick days lost and productivity decrements due to indoor air pollution were in the vicinity of half a billion dollars per year for the United States.

There exists some field study evidence linking productivity decrements to SBS. In their survey of 7000 Dutch workers in 61 office buildings, Preller et al. (1990) found a substantial volume of sick leave attributed to SBS-type symptoms. Raw's re-analysis of the productivity questions used in the "Office Environment Survey" (Wilson and Hedge 1987) yielded a significant linear dependence of office workers' self-assessed productivity on the tally of their reported SBS symptoms (Raw et al. 1990). So, although it may not be possible to quantitatively assess productivity impacts, there would appear to be some empirical support for assigning at least a relative decrease in the output of office workers suffering from SBS.

2 Physical, Chemical and Biological Hypotheses of SBS

Despite the syndrome's relatively short period of official recognition, hypotheses about what actually causes SBS are legion. Unfortunately, the overwhelming majority of these have fairly weak scientific bases, but this is more due to the non-specific nature of the symptom cluster and the vague nature of its building-relatedness than the actual quality of the research effort to date. Since ethical considerations rule out a more rigorous and direct experimental approach to the problem, future epidemiological research will inevitably continue in the speculative mode. However, as the SBS database continues to expand, so too will the list of checkpoints for hypothesis tests enlarge, and the chances of a useful solution to the SBS riddle improve.

Admittedly there is something illogical about searching for a specific environmental cause of SBS, because, if the search is sucessful, then the problem ceases to be classified as SBS and reverts to BRI status. Nevertheless, prominent on the list of possible direct causes are the following (see Raw 1992 for a detailed literature review of the various SBS hypotheses put forward to date):

- Inadequate fresh air ventilation
- Chemical pollutants
- Bioaerosols
- Inorganic dust and other airborne particulates

In addition to these direct causes, several risk or susceptibility factors have also been identified in the larger SBS surveys (e.g. Pickering et al. 1984; Wilson and Hedge 1987). For example, it has been noted that the prevalence of SBS symptoms is consistently higher in air-conditioned buildings than in naturally or mechanically ventilated ones (e.g. Zweers et al. 1992; Fisk et al. 1993), and that certain personal characteristics such as being female or an asthmatic are also associated with higher SBS symptom prevalence. In addition, psychosocial factors such as low job status, low job statisfaction, open plan office layout or low levels of privacy, having a public sector employer, and low levels of perceived control over the indoor climate are typically associated with a higher prevalence of SBS symptoms.

In his exhaustive literature review of possible SBS causes, Raw (1992) concluded that there is evidence against virutally every one of the hypotheses listed above, and that it is extremely unlikely that a single causal mechanism for the syndrome will ever be found. Because of the diverse nature of the symptoms comprising the syndrome, different components of SBS may well turn out to have different causes. This contention is borne out by the observation in most of the large SBS surveys such as Wilson and Hedges' (1987) that different symptoms or symptom clusters within the SBS definition have widely divergent incidences between buildings and between groups of building occupants, suggesting that they may each have their own underlying causal mechanisms.

2.1 Inadequate Ventilation

Since the problem of SBS emerged in the early 1970s when the ratio of fresh outdoor air to recirculated air within air-conditioning systems was being wound back in the name of energy conservation, it is not surprising that ventilation has emerged as one of the most obvious suspects in the SBS mystery. However, the evidence in support of these suspicions is patchy at best (Raw 1992).

With so many potentially confounding variables in the SBS problem, simple comparisons between buildings with varying outdoor air ventilation rates is extremely unlikely to yield a consistent result. Despite this difficulty with field-based research, Mendell (1993) reviewed some of the better-designed field investigations and concluded that, in the air-conditioned building studies, there was a tendency for an increase in the incidence of SBS symptoms in situations with ventilation rates less than the 10 l/s per person currently mandated in the ASHRAE Standard 62 (ASHRAE 1989). Those papers failing to find this association were discounted by Mendell (1993) on the grounds of lack of precision in determining ventilation rates or ventilation efficiency.

The Canadian case study by Donnini et al. (1990) clearly illustrates the complexity of the relationship between ventilation rate and the incidence of SBS symptoms in office buildings. Increasing the amount of outdoor air was found to decrease indoor air concentrations of CO_2, formaldehyde and toluene, and yet the frequency of indoor air quality complaints actually increased significantly. In a Swedish study, Wyon (1992) failed to find any statistical association between the frequency of self-reported SBS symptoms in a hospital ward and the outdoor fresh air ventilation rates which were varied up and down by 30%.

In one of the more rigorous double-blind investigations into the role of ventilation rates in the SBS problem, Jaakkola et al. (1994) rotated 75 office workers between high and low ventilation conditions (21 l/s per person and 6 l/s per person respectively) on a weekly basis. There was no significant difference between the SBS prevalence rates or the perceived indoor air quality between the two ventilation rates, leading the authors to conclude that up to 70% recirculated air can be used in office buildings without adverse effects.

Menzies et al. (1993) performed a similar double-blind cross-over experiment in four office buildings with a total of 1546 subjects. Outdoor ventilation rates were manipulated between 10 and 25 l/s per person on a weekly cycle lasting 6 weeks in total. Results were obtained from the building occupants with a standardised questionnaire on SBS symptoms and perception of office environment. The changes in ventilation levels were not associated with changes in participants' ratings of the office environment or in SBS symptom frequency (crude odds ratio was 1.0). Even after adjusting the logistic regression models for temperature, humidity and air speed effects, the odds ratios still came out to unity.

Indeed, it appears that the stronger the internal validity of the research design, the weaker the support for the hypothesis that inadequate ventilation actually causes SBS. However, there are still further grounds for doubting the ventilation hypothesis. Naturally ventilated buildings typically have lower ventilation rates than their air-conditioned counterparts, during winter at least (Raw 1992), and yet the latter are almost universally found to have higher SBS symptom prevalences. Even Mendell's literature review of some of the larger SBS field surveys (1993) found a consistently higher symptom prevalence in the air-conditioned as opposed to naturally ventilated buildings, notwithstanding that author's conclusion about the significance of 10 l/s per person mentioned earlier.

2.2 Specific Pollutants in Indoor Air

If gross fresh air intake rate per occupant fails to correlate with symptom incidence in well-designed SBS experiments, possibly the fault lies in the presumption that the source of indoor air pollution is exclusively the building occupant? The prevalence of this belief is evidenced by the fact that ventilation rates are actually specified as volumes of outdoor air per unit of time per person. Even the ASHRAE ventilation standard for acceptable indoor air quality (Standard 62–89) specifies a minimum of 10 l/s per person (ASHRAE 1989), despite the abundance of research evidence pointing to alternative indoor air pollution sources. For example, Fanger's extensive IAQ research program in Denmark has clearly established that interior building and furnishing materials, paint, office materials and equipment, even the HVAC system itself, may all contribute to the total load of pollutants in the indoor atmospheric environment of an office building (Fanger 1988; 1990). Therefore, it is not surprising that considerable reserch effort has been directed at linking specific chemical and biological indoor air pollutants and SBS symptoms. The following paragraphs are too few to provide a comprehensive review of the rapidly expanding literature on this issue, but they can indicate some dominant themes emerging from recent SBS research.

2.2.1 Volatile Organic Compounds

Modern building materials such as particle board binders and flooring adhesives, carpet backing, paints and furniture, all off-gas a diverse cocktail of pollutants, but it is the volatile organic compounds (VOC) that have attracted most SBS research attention to date. VOC emissions result from the boiling of compounds at room temperature, so the rate of off-gassing decays exponentially with the age of the material. It is now widely accepted that the interior of a typical modern office building may emit literally hundreds of different VOCs into the internal atmosphere (Mølhave 1985), causing their atmospheric concentrations to greatly exceed outdoor levels.

Unfortunately, application of the industrial hygienists' traditional tools of threshold limiting values (TLV) and dose-response models to the context of SBS symptoms and specific VOCs was doomed from the outset. The reason is simply that the number of chemicals potentially involved in non-industrial indoor climates is too large and their concentrations several orders of magnitude smaller than those typically investigated by industrial hygienists – equipment is just not capable of measuring most them at the required accuracy (Fanger 1988, 1990). However, their total or combined concentrations (TVOC) may quite easily reach significant levels, particularly in the early phases (6–12 months) of an office building's occupancy. But implicit in the use of the TVOC concept as an exposure index is the equal weighting and additivity of its individual components (Mølhave 1995). Nevertheless, European research and regulatory effort is moving towards the classification of indoor climates into three categories based on expected levels of sensory, particularly occular, irritation from TVOC. The limiting value for "no unacceptable prevalance" is estimated to be 0.2 mg/m^3 while exceeding 3.0 mg/m^3 is deemed likely to produce unacceptable discomfort (Mølhave 1995). The middle range between 0.2 and 3.0 mg/m^3 TVOC may possibly be associated with irritation and discomfort in the presence of other interacting exposures.

Probably the only exception to this generalisation about sub-threshold concentrations of individual VOCs is formaldehyde. It achieved infamy during the 1970s when urea formaldehyde foam insulation (UFFI) was widely installed in residences, and then subsequently removed at great expense once the deleterious health and sensory effects of formaldehyde were more fully documented. UFFI is no longer installed in buildings, and since the US-EPA Waterside Mall case, progress has been made towards reducing formaldehyde emissions from carpets. Nevertheless, formaldehyde may still be found in the indoor atmospheric environment of new office buildings due to its off-gassing from certain classes of particle board, tiling adhesive and furniture.

The health effects resulting from exposure to VOCs such as formaldehyde fit the classic SBS picture described above, including mild irritation of the eyes, mucous membranes and upper respiratory tract at levels up to 0.3 mg/m^3 and headaches at higher concentrations (Samet et al. 1987, 1988; Mølhave 1995). However, since these effects can logically be expected to diminish as the VOC concentrations themselves decay throughout the first year of an office building's occupancy, it seems highly unlikely that this class of pollutant can be implicated in the aetiology of the SBS, except perhaps, the temporary SBS variant (Raw 1992). Scepticism is further reinforced by the observation that 1970s office buildings typically have a higher SBS symptom prevalence than those constructed more recently (e.g. Hedge et al. 1989).

2.2.2 Bioaerosols

Airborne biological matter commonly found in indoor atmospheres includes pollen grains (10–50 μm), spores, moulds, viruses and bacteria, dust mites and their faeces. Their sources are outdoor air, soil, HVAC plant (especially humidifiers and cooling towers), and the building occupants themselves. Their health effects range from hayfever-type symptoms in the case of pollens, asthma in the case of house dust mites, through to influenza for viral and bacterial microorganisms. Although these factors should strictly be categorised as BRIs, there exists a sizeable body of research literature implicating them in the SBS problem. For example, Miller and Nathanson (1993) claim that microbial contamination was implicated as the primary cause of poor air quality, and by implication, SBS problems in 20% of the Canadian office buildings they investigated on behalf of Public Works Canada.

In a comparison of the indoor environments in a sample of naturally ventilated and air-conditioned office buildings, Harrison et al. (1990) found generally lower concentrations of fungi and bacteria in the air-conditioned buildings. However, the incidence of SBS symptoms was actually lower among the occupants of naturally ventilated buildings in the study. After statistically correcting for the effects of building ventilation systems (air-conditioned versus naturally ventilated), there was a moderately positive correlation between symptom and bacteria/fungi frequencies, but the majority of variance in symptom frequency was accounted for by the air-conditioned versus naturally ventilated factor.

There is a well-established linkage between the population of house dust mites, or more specifically, their excreta, which acts as an allergen, and the frequency and intensity of asthmatic episodes experienced by building occupants (e.g. Andersen and Korsgaard 1984). Raw et al. (1993) examined the possibility of alleviating SBS symptoms in office buildings by removing dust mites and other pollutants from the office environment with various cleaning regimes. Their two test treatments included cool shampoo and steam cleaning/hot water extraction applied to all soft surfaces (carpets, furniture, curtains) followed up with high efficiency filtration vacuuming and wet wiping, dusting and mite treatment. Relative to the control (placebo), both cool shampoo and steam clean/hot water extraction reduced symptom incidence according to a t-test on before-and-after questionnaire responses.

The steam cleaning treatment was also found to reduce the population of dust mites, leading Raw (1992; Raw et al. 1993) to speculate that these organisms and the their faeces were possibly responsible for the symptoms. However, it is generally accepted that dust mite populations are orders of magnitude higher in homes than workplace settings, in Australia at least. Therefore the office buildings-relatedness of SBS symptoms would seem to eliminates dust mites from the list of likely causes.

2.2.3 Inorganic Dust and Other Particulates in Indoor Air

Raw points out that the evidence supporting airborne dust and particulates as a causal factor in SBS is contradictory (1992). In a cross-sectional survey of three Washington DC office buildings, Wallace et al. (1991, 1993) found that airborne dust count was the physical environmental variable most consistently correlated with questionnaire symptom frequencies. However, given that dust levels in naturally ventilated buildings are typically much higher than those in buildings with HVAC systems, while symptom prevalence is typically higher in the latter, inorganic dust and other airborne particulates in indoor air must logically rate as improbable causal agents for SBS (Harrison et al. 1990).

3 Psychosocial Dimensions of SBS

Apart from the various physical, chemical and microbial hypotheses about what actually causes SBS, the literature indicates that several psychological, social (managerial) and behavioural factors have also featured in various research hypotheses.

3.1 Are SBS Symptoms Real?

Despite the absence of a systematic and consistent picture of causation, it would be wrong to hastily dismiss SBS as an imaginary problem (Anonymus 1991) or ascribe it to mass hysteria (Bauer et al. 1992). Many researchers have clinically validated some of the SBS symptoms elcited by questionnaire. For example, in a comparison of 169 Danish office workers with 112 subjects randomly selected from the general population, Franck et al. (1993) established that objectively diagnosed SBS eye manifestations such as premature break-up of precorneal tear film, absence of foam at the inner eye canthus, and epithelial damage of the conjunctiva were all significantly correlated with self-reported eye complaints, and most frequently recorded from the group of subjects working in buildings with a high incidence of SBS complaints. Unfortunately, what was not made clear in the Danish study was whether it was the office environment per se, the nature of the work conducted therein, or even the character of the organisational culture, that was actually causing the eye symptoms.

3.2 The Plethora of SBS Risk Factors

More commonly regarded as risk factors than actual causes of the SBS certain personal or psychosocial factors have been consistently overrepresented in the symptomatic occupants of so-called sick buildings. Women in clerical positions involving repetitive or routine tasks, having low job satisfaction, being accommodated in open-plan offices, required to operate a computer and handle carbonless photocopy paper, being a long way from a window and having a perception of little or no control over their personal work environment are generally found to be more susceptible to SBS than males (e.g. Skov et al. 1989; Hedge et al. 1989; Zweers et al. 1992; Fisk et al. 1993; Mendell 1993).

Obviously many of these risk or vulnerability factors are highly interconnected. For example, despite recent shifts in the gender balance of institutional structures in the West, women are still more likely than men to hold clerical as apposed to managerial positions. Low-status employees are more likely to be required to share an office with a large number of other employees. Open-plan settings usually entail fewer degrees of freedom to thermoregulate or personalise one's workstation environment to individual requirements and tastes. In short, gender well may be acting as a surrogate for heightened workplace psychosocial stressors (Ryan and Morrow 1992).

One critical shortcoming of this gender hypothesis, however, concerns the much higher symptom prevalences typically found in buildings with centralised HVAC systems compared to naturally ventilated buildings. What gender has to do with the type of ventilation system a building has, either directly or indirectly, is a little difficult to explain. So, while gender may well be a potential root cause in this cluster of personal or organisational risk factors, it is found wanting when it comes to the "architectural and engineering risk factors" of SBS.

In an attempt to disentangle this plethora of risk factors in the aetiology of SBS to date, Hedge et al. 1989 reanalysed the vast database collected during the UK's Office Environment Survey (Wilson and Hedge 1987). The statistical method used was path analysis, and consisted of exploring the pattern of correlations between three risk factor clusters (individual, organisational and architectural), and a personal symptom index derived as the sum of each respondent's SBS symptom reports. Of particular interest to the present chapter is the inclusion of various mediating factors such as perceived environmental control, job stress and environmental satisfaction. Figure 1 depicts the causal pathways found to have statistical significance. The usual caveat concerning multicollinearity in datasets such the British study's should be borne in mind when interpreting Fig. 1.

3.3 Perceived Control Over the Workstation Environment

An alternative to the gender hypothesis that includes architectural and engineering risk factors such as those shown in Fig. 1 is the degree of environmental control, actual or perceived, that a worker has in the workplace. It is quite conceivable that workers in low-status jobs, more probably female than male, are required to work in large open-plan spaces where environmental controls such as windows, blinds and doors are largely nonexistent or inaccessible. Furthermore, being relatively low in status means their access to the building's

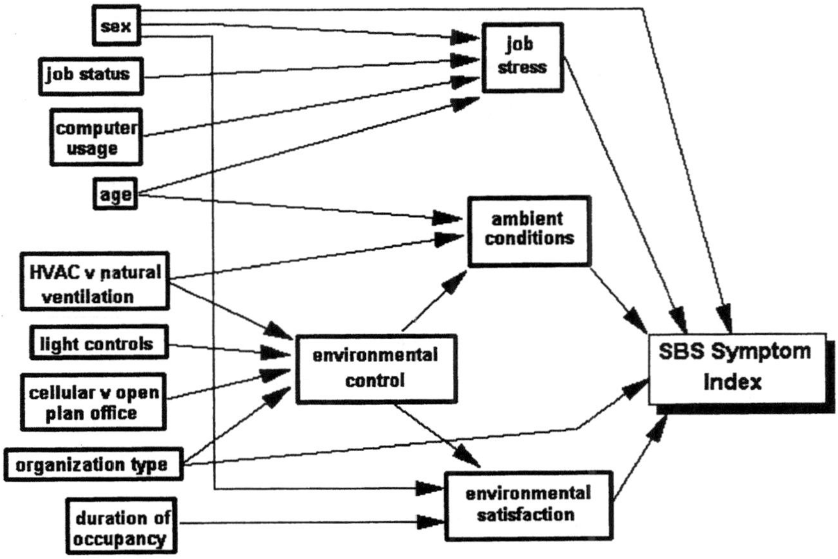

Fig. 1. Path model of the sick building syndrome. (After Hedge et al. 1989)

control (thermostat) and management systems is also minimal, or at least perceived as such.

As indicated in Fig. 1 (Hedge et al. 1989), the Environmental Control factor mediates the effects of ventilation system type (e.g. air-conditioned versus naturally ventilated) and the type of office space (number of workers sharing the space). A total of 45% of variance in the Perceived Control index was accounted for by the path linkages depicted in Fig. 1, and the effects of this perception were transmitted directly downstream to the level of Environmental Satisfaction which, in turn, produced the single largest effect on the Personal Symptom Index (Hedge et al. 1989).

There exists clear evidence linking the number of other people a building occupant is required to share an office with, to that occupant's level of perceived workstation environmental control. Bordass et al. (1993) have conducted questionnaire surveys in several dozen UK office buildings, asking respondents to assess their level of personal environmental control on a seven-point scale starting from 1 = no control through 7 = full control. Average results from all buildings in their cumulative database are presented in Fig. 2. Perceived control clearly diminishes as the office layout changes from single-occupant cellular design to multi-occupant open plan. The exception to this generalisation seems to be the occupant of an open-plan office who is fortunate enough to be sitting next to a window (Bordass et al. 1993).

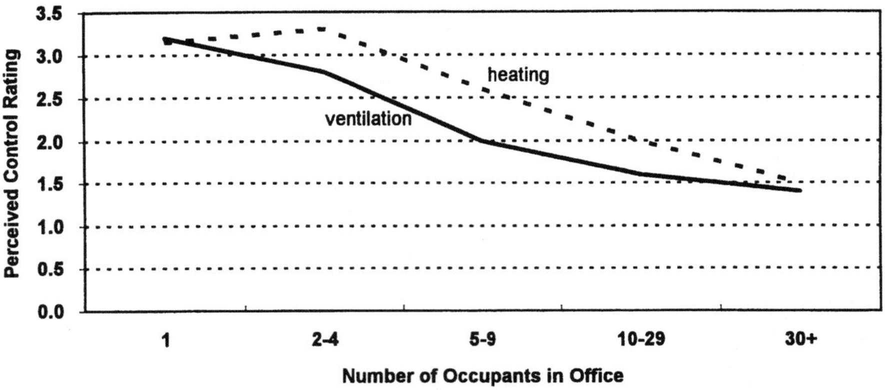

Fig. 2. Loss of environmental control in shared office spaces. (After Bordass et al. 1993)

3.3.1 Stress as a Mediating Variable for SBS

Whilst perceived control turned out to have its largest influence on symptom reporting frequency via environmental satisfaction ratings in the Hedge et al. (1989) model depicted above in Fig. 1, there may well be another pathway not explored in the British Office Environment Survey; namely, environmental stress. Buildings in which the internal environment is centrally and rigidly controlled (i.e. the "zoo-keeper" approach to building energy management) may actually distress their occupants to varying degrees, depending on the individual's vulnerability. Bordass et al. (1933) recently speculated that such no- or slow-response control systems, including architectural (e.g. fenestration), engineering (HVAC) and managerial (facilities managers), are powerful stressors for building occupants, and this has been at least partially confirmed by Bauer et al. (1992) with the finding that self-reported measures of psychopathology and stress on the Minnesota Multiphasic Personality Inventory and Symptom Checklist 90 were higher for occupants of a "sick building" compared to a control sample drawn from the same organization but accommodated in "normal" offices.

The role of perceived control in mediating environmental stressors has long been accepted in environmental psychology. For example, studies on noise have indicated that perceived control significantly reduces both stress and adverse aftereffects of exposure (Glass and Singer 1972).

If the possibility is conceded that remotely controlled environments can distress their occupants, the next question must be "Can this psychological stress actually induce symptoms of the SBS variety?" Considerable anecdotal evidence liking stress with immune dysfunction is supported by some epidemiological evidence. For example, Meyer and Haggerty (cited by Husband 1990) report that acute or chronic family stress was associated with upper respiratory streptococcal infection. Certain other forms of stress such as the occupational variety that results from workers' inability to personally control work demands have also been implicated in adverse hormonal stress reactions (Frankenhauser and Johnsson 1986) along with a battery of symptoms not dissimilar to those partially characterising SBS – viz, headache, fatigue and difficulty in concentrating (Hedge et al. 1989; see Fig. 1).

In addition to these epidemiological studies, there is also some clinical evidence for stress-induced immunosuppression. For example, Husband (1990) reviewed research data showing that lymphocyte proliferation rates were reduced in subjects following recent bereavement. Lymphocytes represent the "smart" component of the immune system, capable of recognising antigens and focusing antibody activity specifically at individual foreign substances, and then retaining an immunologic "memory" of the event for future reference and rapid depolyment. In a prospective clinical study, Jemmott et al. (1983, cited by Husband 1990) found significant negative correlations between salivary antibody counts and academic stress levels experienced by students of dentistry before and after examinations.

As to the hormonal mechanisms underlying stress-induced immunosuppression, stressors are known to stimulate the pituitary gland into triggering the production of corticosteroids by the adrenal gland. Corticosteroids are recognised as powerful immunosuppressive agents and have been used for many years in clinical practice to inhibit unwanted immune responses such as allergies, inflammation and autoimmune disorders (Husband 1990).

3.3.2 Perceived Environmental Control, SBS and Productivity

Apart from carrying implications for SBS symptomatology, workplace stress and low levels of perceived environmental control may also impact on employee productivity. That is, a "productivity" box might well be inserted into the path model of SBS (Fig. 1 above), immediately downstream from "SBS Symptom Index". Whether this impact stems directly from perceived environmental control or is mediated via SBS symptom prevalence remains unclear at this stage (Wilson and Hedge 1987); however, in a recent survey of 1476 workers in 58 office buildings within the same UK organisation, Bordass et al. (1993) found significant correlations between office workers' perceived level of control (even point scale) and their self-assessed productivity (on a 9-point scale with 1 = 40% decrement, 5 = no effect, 9 = 40% gain). Perceived control over heating had a Spearman's correlation coefficient of 0.62 (p = 0.003), perceived control over cooling was found to have a Spearman's correlation coefficient of 0.67 (p = 0.001), while perceived control over ventilation had a correlation of 0.47 (p = 0.02) with productivity (Bordass et al. 1993). Whether subjects who are psychologically predisposed to respond "yes" to SBS symptom checklist items are also predisposed to rate themselves as low-productivity employees is a moot point and virtually impossible to assess with simple questionnaires of the type used by Wilson and Hedge (1987), but the clear linear negative relationship

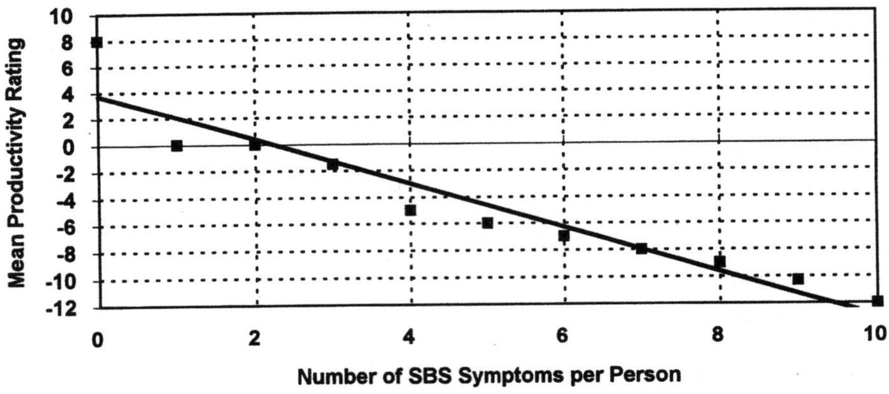

Fig. 3. Impact of SBS symptom prevalence on self-ratings of productivity by office workers. (After Wilson and Hedge 1987)

depicted in Fig. 3 is not inconsistent with the hypothesis being developed in the present chapter.

4 Appropriate Design for the Avoidance of SBS: Task Conditioning

While a naturally ventilated office environment may well be the most habitable design, it seems almost inevitable that the deep-space-plan office building is here to stay, largely due to site constraints in most central business districts of cities around the world. With deep-space planning comes the need for HVAC services, usually to the exclusion of operable windows, but this need not force indoor climate controls into the centralised "zoo-keeper" mode. Raised floor air distribution designs are now becoming increasingly common in open-plan offices for a variety of reasons such as greater flexibility in configuring the layout of electrical and communication networks. A raised floor also provides the opportunity for individually controlled workstation microclimates, since the point of delivery from the underfloor plenum can be directed into the occupied zone of each and every workstation, and reconfigured as required by periodic reorganisation of the office space. This design of air conditioning has been called "task conditioning", and, as with task lighting systems, the settings for these systems can be placed partially or totally under the control of the occupants (Heinemeier et al. 1990). Other terms also in use include "localised thermal distribution systems" (LTDS), "personalised air-conditioning" (PEC) and "environmentally responsive workstations" (ERW) (Kronor and Stark-Martin 1994). An ERW integrates not only heating, cooling and ventilation, but also lighting, acoustic and other environmental parameters directly under the control of the workstation's occupant.

4.1 Design Strategies of Personal Enivronmental Control

Task conditioning requires the delivery of conditioned air to individual workstations. The typical method has involved raised floors that create an underfloor plenum, or at least a space which is large enough to conceal flexible ducts to each workstation. Three broad types of task-conditioning terminal units have emerged in the market to date (Bauman et al. 1993). With floor-based systems air is either drawn from an underfloor plenum or ducts with user-controlled variable speed fans, or forced through the underfloor plenum and delivered to individual workstations by floor-level supply grilles. Desk-based systems typically supply conditioned air in variable volume through grilles on the back of the desk or through small directable nozzles mounted on the rear corners of the desk's working surface. Another desk-based design delivers low-speed conditioned air to the occupied zone with fan-assisted air handling units mounted in front of the occupant's knees. The third style of task conditioning, popular in Japan, supplies conditioned air through grilles mounted in open-plan office partition panelling.

In all three designs, supplementary microclimatic control is often provided by devices such als low-wattage radiant panels mounted either on the underside of the desk or on the office partition panels beside the workstation occupant. Furthermore, additional cleaning of supply air just prior to its delivery into the occupant's breathing zone can be achieved in all three classes of task conditioning by the incorporation of mechanical and electrostatic filtration devices within the fan-assisted outlets, thereby enhancing the ventilation efficiency of task-conditioning systems in comparison to more traditional ambient conditioning strategies.

A recently completed "smart" office building in Tokyo, as described by Matsunawa et al. (1995), provides a suitable case study of the task-conditioning design concept. The eight-storey building was completed in 1992 and provides 1200 m^2 floor area on each level for each of its two adjacent wings. A central atrium separates the two wings with no partitions between office areas and atrium across the full eight storeys. Outdoor air inlets are provided in he spandrel panels on each floor and are remotely operable to provide additional ventilation of the underfloor HVAC plenum whenever outdoor weather and seasonal conditions are close to indoor requirements, thereby enhancing energy conservation potential. Air pressure within the 30-cm-deep underfloor plenum is up to 10 Pa higher than in the occupied zone, and conditioned air is supplied indoors via floor-mounted diffusers spaced approximately one per 10 m^2, but which may be relocated easily to cope with changing amounts and locations of internal loads. Air is returnred to the HVAC plant via slit-type inlets mounted in the ceiling light fixtures. The floor mounted air diffusers were designed to provide different jet patterns depending on whether the HVAC is operating in heating or cooling mode. In heating mode, the diffusers project air outwards across a wide floor area, while in cooling mode, they project a vertical jet up into the occupied zone. The diffusers are fan-assisted and this feature, along with jet direction, confers some localised control on the system via a small infrared hand-set. The fan settings across the entire floor area of a single storey are constantly monitored by a centralised air handling unit which enables the temperature of the supply air within each underfloor plenum to be automatically adjusted to meet total demands from that particular floor's occupants. Supplementary localised control is provided to each occupant by ducting conditioned air from the same fan-assisted floor diffusers into open-plan office partitioning and other furniture, terminating in desk-height diffusers which direct the air into the workstation's occupied zone. An additional task-conditioning feature of the building consists of moderate temperature radiant panels in the under-desk region of each workstation, to counter any draft risks posed by higher air speeds at these levels.

4.2 Advantages of Task Conditioning

Several advantages of task conditioning are obvious, particularly those dealing with spatial and temporal variations of environmental requirements within the building. Spatial differences arise from interindividual variations in the thermal and ventilation requirements of the building occupants. These can be caused by variations in physical exertion, clothing patterns, stress levels, along with heat loads such as computers and printers on desks. Experimental evidence indicates that interindividual differences in the subjective comfort setpoint alone can amount to more than 1 °C (Fanger 1970). Preferences in relation to air movement and ventilation can also vary significantly from one person to the next (Fountain et al. 1994). Furthermore, intraindividual fluctuations in any of these factors during the course of the day can also occur, and task conditioning with individual control devolved to the level of the individual workstation is capable of handling these as well.

In view of the variety of factors causing divergent end-user requirements of the office environment, it should come as no surprise that recent field investigations of office thermal comfort in buildings with conventional HVAC systems (e.g. Schiller et al. 1988; Schiller 1990; de Dear and Fountain 1994) have uncovered dismally low levels of overall thermal satisfaction among workers, even in environments conforming to temperature, humidity and air speed prescriptions set down in the indoor climate standards such as ASHRAE 55-92 (1992) and ISO 7730 (1984). Certainly the HVAC engineer's ideal target, as set out in Fanger's (1970) classic text on thermal comfort, of satisfying 95% of the occupants within a single space at any one time is exactly that – an ideal which, regrettably, appears unattainable with conventional approaches.

4.2.1 Task Conditioning and SBS

Task conditioning is generally assumed to minimise the incidence of SBS symptoms due to its improved ventilation efficiency. This efficiency stems from the delivery of relatively fresh-supply air directly into the breathing zone of each building occupant, instead of coming into the room from a ceiling diffuser and being mixed with all the pollutants in that room before eventually being inhaled by the occupants (Bauman et al. 1993). However, another SBS benefit may well be the enhanced controllability of workstation microclimate. As discussed above, one of the clearest correlates of the SBS is the perception of lack of control over the physical or work environment (Raw 1992). Currently available evidence does not clearly establish whether this is because controllability is an important variable in its own right, or if individual control gives the ability to individuals to create an environment which suits them" (Raw 1992). Pragmatically, this distinction is irrelevant - task conditioning certainly confers controllability on the individual workstation occupant, and, in all probability, will reduce his or her susceptibility to SBS-type symptoms. As Linvall (1985) pointed out, perceived discomfort is a critical effect due to the mental nature of office work. Even when the physical parameters depart from an individual's preference by only a small amount, they can still be psychologically distressing, thereby increasing the risk of cumulative effects and either making the worker more susceptible to minor SBS-type ailments, or making them more prone to complain about HVAC services and indoor air quality.

Statistical evidence for the link between control and satisfaction with air quality, if not the incidence of SBS symptoms directly, was provided by O'Neill's (1992) field study in a US office building. The study sample consisted of 200 office workers on five different floors of a single building that used either task-conditioning air distribution units or a standard, zone-controlled HVAC system. The task conditioning units were connected to an under-floor air distribution plenum which permitted individual control over both volume and direction of airflow within each workstation. Dependent measures of air quality included perceptions of (a) air freshness, (b) temperature, (c) seasonal variations in temperature, and (d) satisfaction with temperature. The statistical analysis consisted of multiple regression equations in which sex and occupational status were controlled. The results indicated that, after sex and job status effects were statistically controlled, personal control was a significant predictor of employee perceptions and satisfaction. In order of importance: exercised control, perceived control, the importance of air quality as a goal, physical enclosure, and availability of control all significantly and postively influenced perceptions of thermal comfort and satisfaction.

4.2.2 Task Conditioning and Employee Productivity

Even if the evidence linking task conditioning to reduced SBS symptom prevalence in offices is sparse to date, there are several other advantages which may independently justify the added expense of such designs. Until firmer SBS data are available, the SBS goal can be "piggy backed" on the following, more tangible benefits of environmentally responsive workstations.

One potentially attractive aspect of task conditioning is the suggestion that it may enhance employee productivity. There exists already some field-based evidence showing this benefit of localised control. Kronor and Stark-Martin (1994) used an index of employee productivity developed in-house by the West Bend Insurance Company, Wisconsin, to quantify the effects of that corporation's move into premises fitted out with 370 ERW's. The research design was a simple before-and-after statistical analysis of employees individual productivity data, and as such, the ERW effect was confounded by the change in building. To counter this research design flaw, Kronor and Stark-Martin (1994) devised a placebo in the form of disabled temperature, air speed and radiant heater panel functions in a randomly selected sample of ERWs, without informing the subjects. Also, because the underwriters' productivity questionnaires were routinely collected by their employer even before this study, the subjects were naive to the purpose of the study, being told that some ERWs might be temporarily disabled to assess their energy consumption.

The largest single factor impacting on productivity in the Wisconsin study was the disruption caused by the move between buildings (estimated to be -32%). However, after statistically adjusting for the effects of the move, the unique contribution of the environmentally responsive workstations to employee productivity was of the order of +2%. Whilst not being a huge productivity gain, saving 2% of its payroll was deemed by the insurance company to more than adequately justify the additional costs of task conditioning over a more conventional ambient air-conditioning design.

4.2.3 Task Conditioning and Energy Conservation

Ambient temperatures within the building can drift closer to the outdoor temperature, as long as the temperature of the air being supplied to the individual workstation is suitable for mixing to the occupant's preference. This has the potential to reduce building energy consumption, since the occupied space within a building is just a fraction of the total enclosed space. Another energy-conservative feature of task conditioning is that each workstation can easily be equipped with an occupancy sensor that powers down the environmental systems whenever the workstation is vacated for any longer than a predetermined interval - typically about 10 min - and then reactivates them with the same settings whenever the occupant returns.

Quantitative assessments of the impacts of personal environmental control on building energy use have yielded encouraging results so far. The Tokyo case-study building described earlier was subjected to post-occupany energy audits (Matsunawa et al. 1995). The system was found to have a significantly higher-than-usual temperature effectiveness index, defined as the temperature gradient between return air and supply air divided by the temperature gradient between occupied zone and supply air. High temperature effectiveness translates into high energy conservation potential, since conditioned air is being delivered directly into the occupied zone of each workstation without mixing and dilution, and also that waste heat generated within the occupied zone by occupants themselves, office equipment and lighting, is being exhausted efficiently without mixing. In the Tokyo case-study building, high temperature effectiveness index values were achieved with supply air temperatures at about 20 °C, indicating considerably reduced cooling requirement compared to more conventional ceiling-based air-conditioning systems, which require supply temperatures down to 16 °C in order to deal with the same loads and provide the same occupied zone temperatures.

Furthermore, the exploitation of as much natural cooling as climatically feasible was achieved in the Tokyo case-study building by means of energy-conscious management of the direct air intakes on each floor. Having a supply air temperature at least 4 °C warmer than typical for conventional systems meant that the duration of the natural cooling seasons in Tokyo's climate was greatly extended. The net effect of this air-side economizer feature was to reduce cooling energy consumption in the transitional seasons by as much as 30% below levels normally expected for conventional HVAC designs. Furthermore, higher supply air temperatures permitted higher chilled water temperatures within the refrigeration plant (up from 5 to 9 °C), which, in turn, almost doubled the refrigeration coefficient of performance (COP up from 3.6 to 6). Overall annual energy consumption by the building's task-conditioning HVAC system per unit floor area was audited to be 24% lower than comparable Tokyo office buildings with conventional ceiling-based air-conditioning systems, clearly refuting the often-stated assumption that devolution of air-conditioning control to individual building occupants is synonymous with increased energy consumption.

5 Conclusions

Sick buildings are those in which a sizeable group of occupants report a diverse array of non-specific symptoms which either disappear completely or are significantly relieved upon leaving the building. The symptoms appear more frequently in particular buildings or are concentrated in specific parts of them. The symptoms include sensory irritation, neurologic or general health symptoms, skin irritation, nonspecific hypersensitivity reactions and odour and taste problems. Irritation of the mucous membranes in eye, nose and throat are among the most frequent symptoms (Mølhave 1992).

Most cases of SBS reported to date have suggested that it is significantly linked with occupancy of air-conditioned, energy-efficient structures with a sealed building envelope. Therefore most investigations have logically searched for causes in specific atmospheric environmental agents which would probably be exacerbated by low fresh-air ventilation rates. Prominent in the search have been bioaerosols, volatile organic compounds and dust/particulates. However, even though specific atmospheric agents such as these have occasionally been associated with temporary sick buildings (usually new or recently remodelled offices), investigations of indoor air pollutants in permanently sick buildings have usually turned up null findings.

What has emerged from the larger SBS investigations is the importance of social and psychological dimensions of the office environment in relation to the actual or perceived health status of its occupants. Factors highly correlated with the incidence of SBS symptoms have included:

- Multi-occupant open-plan office layout
- Gender, with females more at risk than males,
- Job-Status, with clerical and secretarial personnel more at risk than managerial and professional
- Low levels of perceived control over local environment

Obviously SBS is a multifactorial problem, but in the present chapter it has been argued that the last of these psychosocial factors may be confounding several others on this list of risk factors. The degree of environmental control an office worker is capable of exercising seems to be inversely proportional to the number of other employees with whom he or she is required to share the space. Managerial and professional staff are more likely to occupy the private offices or cubicles found on the perimeter of many modern offices plans, and, furthermore, are likely to exert considerably stronger influence (by way of complaint) over the running of a building's heating, ventilation and air-conditioning (HVAC) services than the lower-ranked, more probably female, personnel within the same organization. In contrast, naturally ventilated buildings have operable windows along with a much more diverse set of local indoor climatic control options for their occupants to manipulate. In short, they provide a much more interactive and democratic indoor environment and climate, and have consistently been found in the major SBS symptom prevalence surveys to be healthier, either actually or perceived, compared to conventional centrally controlled air-conditioned counterparts.

While natural ventilation and small, single-occupant offices may well be the most appropriate strategy for the minimisation of SBS, they simply are not practical in most central business district sites. However, this fundamental requirement for building occupants to interact with and regulate their physical environment is not totally incompatible with the modern deep-space planning dictated by the typical office building design brief. In recognition of the diversity of individual human needs, several recent so-called "smart buildings", such as Lloyd's of London and Chiffley Tower in Sydney, have pioneered a new HVAC strategy called "task conditioning". Task conditioning typically uses underfloor air distribution in open-plan office designs, supplying conditioned air to individual workstations, the occupants of which are free to blend it in varying proportion with room air, or electrically reheat it to suit precisely their personal metabolic, clothing and subjective requirements, as well as coping with the unique heat loads impinging on their individual workstation. In devolving some degree of environmental control back to the individual office worker, task conditioning presents a more appropriate HVAC strategy for "healthy buildings" than the conventional, centralised air-conditioning found in the overwhelming majority of "sick buildings".

Acknowledgements. Donna Cooper is thanked for her assitance in searching the literature during the preparation of this chapter. All delegates to the Climatic Impacts Centre's Workshop on the Indoor Atmospheric Environment on 13 May 1994 are also thanked for providing feedback on the ideas expressed in this chapter, but final responsibility for these opinions rests with the author.

References

Andersen L, Korsgaard J (1984) Asthma and the indoor environment: assessment of the health impacts of high humidity. Proc 3rd Int Conf on Indoor Air Quality and Climate, Stockholm V 1:79–88

Anonymous (1991) Editorial: Sick building syndrome. Lancet 338:1493–1494

ASHRAE (1989) Standard 62-89: Ventilation for acceptable indoor air quality. American Society of Heating, Refrigerating and Air-Conditioning Engineers, Atlanta

ASHRAE (1992) Standard 55-92: Thermal environmental conditions for human occupancy. American Society of Heating, Refrigerating and Air-Conditioning Engineers, Atlanta

Axelrad R (1989) Economic implications of indoor air quality and its regulation and control. NATO/CCMS Worksh, Pilot study on Indoor air quality, the implications of indoor air quality for modern society. CCMS Rep No 183, NATO/CCMS, pp 89–116

Bauer RM, Greve KW, Besch E, Schramke CJ, Crouch J, Hicks A, Ware MR, Bradford Lyles W (1992) The role fo psychological factors in the report of building-related symptoms in sick building syndrome. J Consult Clin Psychol 60:213–219

Bauman FE, Zhang H, Arens EA, Benton C (1993) Localized comfort control with a desktop task conditioning system: laboratory and field measurements. ASHRAE Trans 99:733–749

Bordass WT, Bromley AKR, Leaman AJ (1993) User and occupant controls in office buildings. In: Bieva C, Collett C (eds) Proc Int Building design, technology and occupant well-being in temperate climates. ASHRAE, Atlanta, pp 12–17

Colligan MJ, Murphy LR (1979)Mass psychogenic illness in organizations: An overview. J Occup Psychol 52:77–90

de Dear RJ, Fountain ME (1994) Field experiments on occupant comfort and office thermal environments in a hot-humid climate. ASHRAE Trans 100:457–475

Donnini G, Nguyen VH, Haghighat F (1990) Ventilation control and building dynamics by CO_2 measurement. Proc 5th Int Conf on Indoor Air Quality and Climate, Toronto V 4:257–262

Fanger PO (1970) Thermal comfort. Danish Technical Press, Copenhagen

Fanger PO (1988) Introduction of the olf and decipol units to quantify air pollution perceived by humans indoors and outdoors. Energy Buildings 12:1–6

Fanger PO (1990) New principles for a future ventilation standard. Proc 5th Int Conf on Indoor Air Quality and Climate, Toronto V 1:353–363

Finnegan MJ, Pickering CAC, Burge PS (1984) The sick building syndrome: prevalence studies. Br Med J 289:1573–1575

Fisk WJ, Mendell MJ, Daisey JM, Faulkner D, Hodgson M, Nematollahi J, Macher M (1993) Phase I of the California healthy building study: a summary. Indoor Air 3:246–254

Fountain ME, Bauman F, Arens E, Miura K, de Dear R (1994) Locally controlled air movement preferred in warm environments. ASHRAE Trans 100:937–952

Franck C, Bach E, Skov P (1993) Prevalence of objective eye manifestations in people working in office buildings with different prevalences of sick building syndrome compared with the general population. Arch Occup Environ Helath 65:65–69

Frankenhauser M, Johansson G (1986) Stress at work: psychobiological and psychosocial aspects. Int Rev Appl Psychol 35:287–299

Glass DC, Singer JE (1972) Urban stress: experiments on noise and social stressors. Academic Press, New York

Harrison J, Pickering CAC, Faragher EB, Austwick PKC (1990) An investigation of the relationship between microbial and particulate indoor pollution and the sick building syndrome. Proc 5th Int Conf on Indoor Air Quality and Climate, Toronto V 1:149–154

Hedge A, Burge PS, Robertson AS, Wilson S, Harris-Bass J (1989) Work-related illness in offices: a proposed model of the sick building syndrome. Environ Int 15:143–158

Heinemeier KE, Schiller GE, Benton CC (1990) Task conditioning for the workplace: issues and challenges. ASHRAE Trans 96:678–688

Hodgson M (1992) Field studies on the sick building syndrome. Annal NY Acad Sci 641:21–35

Husband A (1990) Stress, behaviour and the immune system. Managing stress at work – Royal Sydney Hospital Occupational Health and Safety Unit Workshop, November 1990, Sydney

ISO (1984) International Standard 7730. Moderate thermal environments – determination of the PMV and PPD indices and specification of the conditions for thermal comfort. International Standards Organization, Geneva

Jaakkola JJK, Tuomaala P, Seppanen O (1994) Air recirculation and sick building syndrome: a blinded crossover trial. Am J Public Health 84:422–428

Jemmot JB, Borysenko JZ, Borysenko M, McLelland DC, Chapman R, Meyer DJ, Benson H (1983) The Lancet i:1400–1402

Kronor W, Stark-Martin (1994) Environmentally responsive workstations and office-worker productivity. ASHRAE Trans 100:

Linvall T (1985) Exposure limits for office environments. Ann Am Conf Occup Hyg 12:99–108

Matsunawa K, Lizuka H, Tanabe S (1995) Development and application of an underfloor air-conditioning system with improved outlets for a “smart” building in Tokyo. ASHRAE Trans 101:

Mendell M (1993) Non-specific symptoms in office-workers: a review and summary of the epidemiologic literature. Indoor Air 3:227–236

Menzies R, Tamblyn R, Farant JP, Hanley J, Nunes F, Tamblyn R (1993) The effect of varying levels of outdoor-air supply on the symptoms of sick building syndrome. Engl J Med 328:821–827

Meyer RJ, Haggerty RJ (1962) J Pediatr 29:539–549 (cited by Husband 1990)

Miller JD, Nathanson T (1993) Microbial contamination in buildings – measurement and guidelines. Seminar/Workshop on Indoor Air Quality in Office Buildings, Public Works Canada, Ottawa, 1–5

Mølhave L (1985) Volatile organic compounds as indoor air pollutants. In: Gammage RB, Kaye SV (eds) Indoor air and human health. Lewis, Michigan

Mølhave L (1987) The sick building: A sub-population among problem buildings? Proc 4th Int Conf on Indoor Air Quality and Climate, Berlin V 2:469–473

Mølhave L (1992) Controlled experiments for studies of the sick building syndrome. Ann NY Acad Sci 641:46–55

Mølhave L (1995) The use of the TVOC concept in source characterization and regulation of IAQ. In: Haghighat F (ed) Indoor Air Quality, Ventilation and Energy Conservation in Buildings, vol I, pp 1–19

Molina C, Pickering CAC, Valbjorn O, de Bortoli M (1989) Sick building syndrome – a practical guide. Commission of the European Communities, COST 613 Report 4. ECSC-EEC-EAEC, Brussels

O’Neill MJ (1992) Effects of workspace design and environmental control on office workers’ perceptions of air quality. Proc Human Factors Society, Santa Monica V 2:890–894

Pickering CAC, Finnegan MJ, Robertson A, Burge S (1984) Sick building syndrome. Proc 3rd Int Conf on Indoor Air Quality and Climate, Stockholm V 3:321–325

Preller L, Zweers T, Brunekreef K, Boleij JSM (1990) Sick leave due to work related health complaints among office workers in the Netherlands. Proc 5th Int Conf on Indoor Air Quality and Climate, Toronto V 1:227–230

Raw GJ (1992) Sick building syndrome: a review of the evidence on causes and solutions. HSE Rep No 42/1992. HMSO, London

Raw GJ, Roys MS, Leaman A (1990) Further findings from the office environment survey: productivity. Proc 5th Int Conf on Indoor Air Quality and Climate, Toronto V 1:231–236

Raw G, Roys MS, Whitehead C (1993) Sick buidling syndrome: cleanliness is next to healthiness. Indoor Air 3:237–245
Rowe D, Wilke S (1994) Sick building syndrome and indoor air quality: perception and reality compared. Proc12th Int Conf of the Clean Air Society of Australia and New Zealand. Promarco Conventions Pty Ltd, Perth, pp 615–622
Ryan CM, Morrow LA (1992) Dysfunctional building or dysfunctional people: an examination of the sick building syndrome and allied disorders. J Consult Clin Psychol 60:220–224
Ryd H (1991) My home is my castle – psychological perspectives on the sick building syndrome. Building Environ 26:87–92
Samet JM, Marbury MC, Spengler JD (1987) Health effects and sources of indoor air pollution. Part I. Am Rev Respir Dis 136:1486–1508
Samet JM, Marbury MC and JD Spengler (1988) Health effects and sources of indoor air pollution. Part II. Am Rev Respir Dis 137:221–242
Schiller GE (1990) A comparison of measured and predicted comfort in office buildings. ASHRAE Trans 96:609–622
Schiller GE, Arens E, Bauman F, Benton C, Fountain M, Doherty T (1988) A field study of thermal environments and comfort in office buildings. ASHRAE Trans 98:280–308
Seem JE, Braun JE (1992) The impact of personal environmental control on building energy control. ASHRAE Trans 98:903–909
Skov P, Valbjørn O, Pedersen BV (1989) Influence of personal characteristics, job related factors and psychosocial factors on the sick building syndrome. Scand J Work Environ Health 15:286–295
Wallace LA, Nelson CJ, Dunteman G (1991) Workplace characteristics associated with health and comfort concerns in three Washington DC office buildings. Proc IAQ'91 – Healthy Buildings (ASHRAE), pp 56–60
Wallace LA, Nelson CJ, Highsmith R, Dunteman G (1993) Association of personal and workplace characteristics with health, comoft and odor: a survey of 3948 office workers in three buildings. Indoor Air 3:193–205
WHO (World Health Organisation) (1983) Indoor air pollutants: exposures and health effects. WHO Regional Office for Europe, Copenhagen
Wilson S, Hedge A (1987) The office environment survey: a study of building sickness. Building Use Studies, London
Woods J (1989) Cost avoidance and productivity in owning and operating buildings. Occup Med: State-of-the-Art Rev 4:753–770
Wyon DP (1992) Sick buildings and the experimental approach. Environ Technol 13:113–122
Zweers T, Preller L, Brunekreef B, Bolij JSM (1992) Health and indoor climate complaints of 7043 office workers in 61 buildings in the Netherlands. Indoor Air 2:127–136

Bioclimatic Architecture and Solar Energy

S. V. SZOKOLAY[1]

1 Introduction

Climatic design of buildings became an issue when, in the post-war years, European and North American architects started working in unusual climates – the tropics. The unfamiliar nature of these climates necessitated an analytical examination. This later produced a feedback effect: an examination also of the ostensibly known climates.

Le Corbusier's work in Chandigarh was on a grand scale; he was driven by his "feel", but at that stage knowledge was badly lacking. He was essentially an artist. His concern for sun-control is witnessed by a series of sketches (Fig. 1), which have an emotive appeal, but are not terribly informative, and his *bris-soleils* (sun-breaks) did not quite work as expected. Housing in the same city designed by Fry and Drew was a much more sound response to the climate. They later produced their book *Tropical architecture* in 1964, which had a significant influence on practitioners.

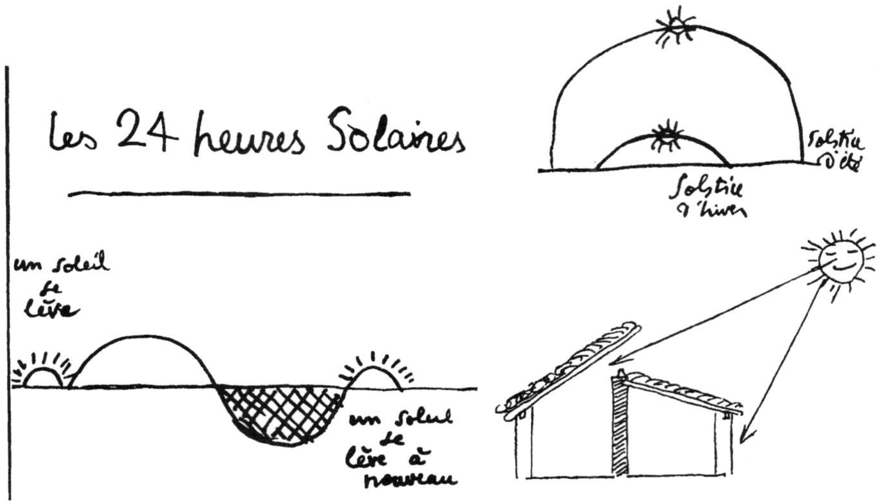

Fig. 1. Some of Le Corbusier's solar sketches (original)

[1] The University of Queensland, Brisbane, Queensland 4072, Australia

This was preceded by a whole series of research projects and publications. D. H. K. Lee (professor of physiology at the University of Queensland) published his *Physiological principles in tropical housing* in 1944. Phillips produced his stereographic solar charts and shading design method in 1948 (some 15 years later also adopted by the UK Building Research Station). The American Institute of Architects published their *Regional climate analyses* in 1949, under the direction of Paul Siple. Drysdale's *Designing houses for Australian climates* was published in 1952.

2 Bioclimatic Architecture

The turning point, in my view, was the paper by the Hungarian-American Victor Olgyay in 1953, entitled *Bioclimatic approach to architecture.* In this the author managed to synthesise relevant products of many different sciences, from physiology through climatology to building physics.

The origin of the word "bioclimatology" (or its adjectival form "bioclimatic") is uncertain, but it was already used in the 1930s, e.g. in the *Bioklimatische Beiblätter* of the *Meterologische Zeitschrift.*

Olgyay devised his *bioclimatic chart* (Fig. 2) to show the effects of four atmospheric variables on human comfort. Dry-bulb temperature is on the vertical axis, relative humidity on the horizontal and a comfort zone is outlined in the middle, in terms of these two variables. Lines above this indicate an extension of the upper comfort limit by air movement and lines below it show a downward extension of the comfort zone by solar radiation. A "dry" version of this chart, converted to SI units, is shown in Fig. 3.

Olgyay coined the term *bioclimatic design,* which became an expression and symbol of the ideology of many architects up to the present day. It is written on the banners of those who oppose the domination of formal concerns, of pseudo-philosophical movements and fashions (post-modernism, deconstruction) in architecture, those who are more concerned with the real problems of the global environment. It expresses an attitude to design which can be summarised in three points:

- The purpose of architecture is the human being.
- (S)he is exposed to a climate, which is not always favourable.
- The task is to create a "filter" between the human and the climate.

The word *filter* is important. Not a "barrier", coupled with energy-hungry heating and cooling, but a selective filter, which excludes undesirable elements (wind, rain, coldness) but admits desirable influences, such as daylight, solar radiation in winter or fresh air. Its result is an architecture which follows the commands of an ecological ethic: take the least resources from and dump the least wastes into the environment.

In the same year (1953) Aronin wrote his *Climate and architecture* book. The *Solar control and shading devices* volume of the Olgyay brothers is a 1957

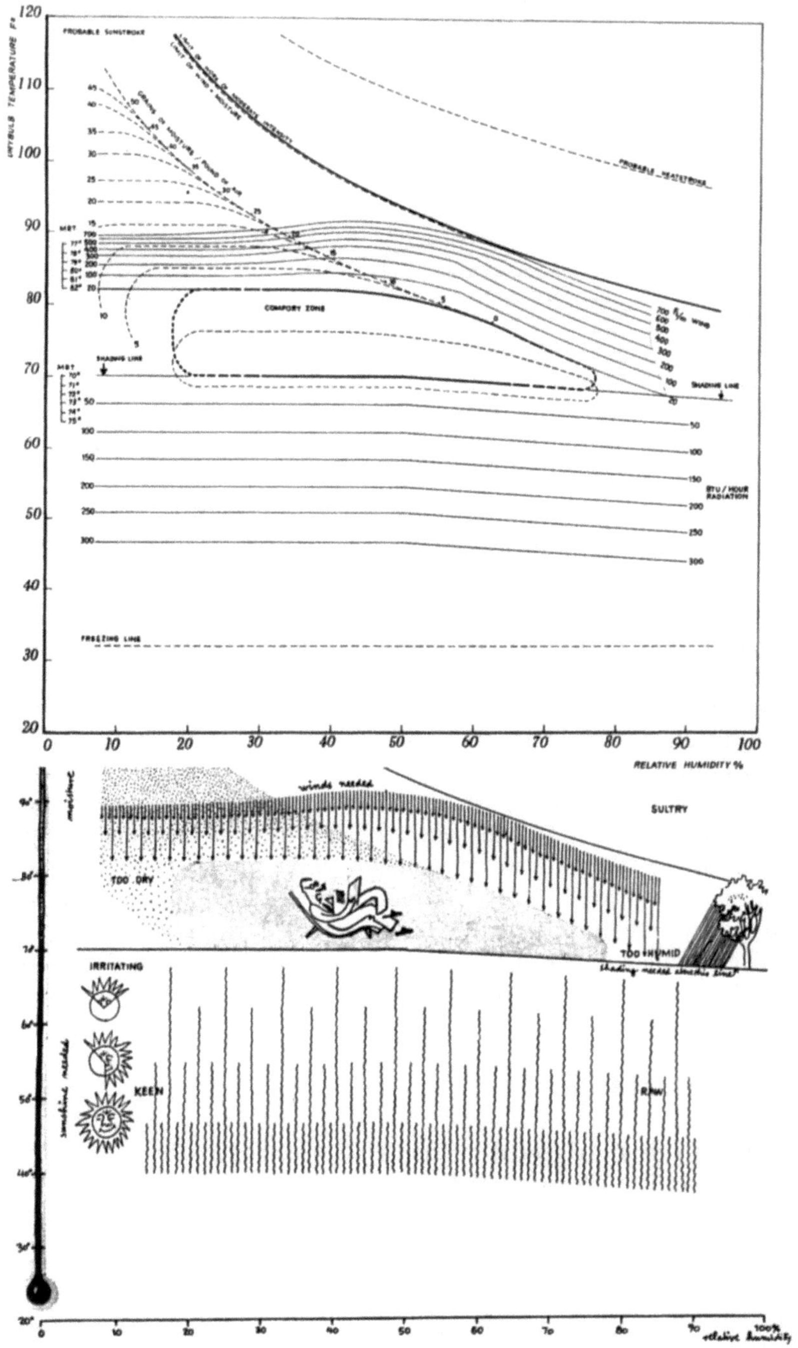

Fig. 2. Olgyay's "bioclimatic chart" and its graphic explantation. The comfort zone is shown for 40° latitude. For every 5° reduction in latitude, increase the temperature should be increased by 1.35 °C

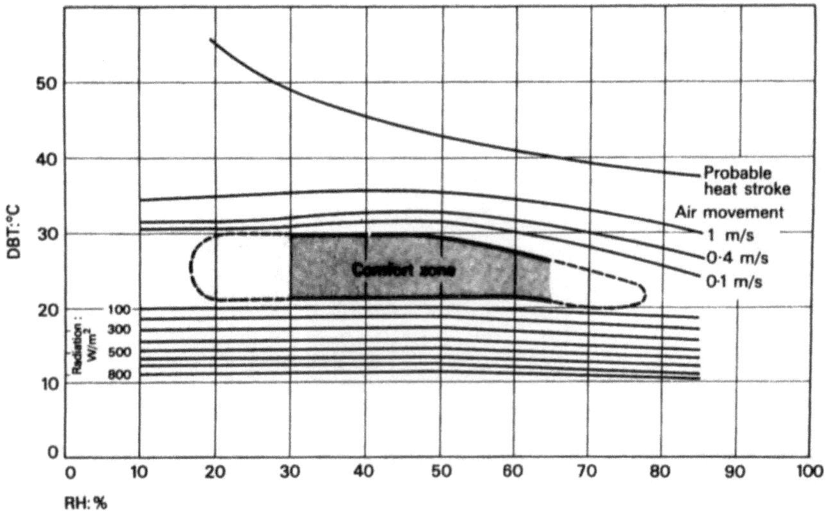

Fig. 3. The same chart as in Fig. 2, converted to SI units and corrected for warm climates

product. Victor's most influential work was the book *Design with climate,* written in 1962, with the subtitle *Bioclimatic approach to architectural regionalism.* The last word here also implied an intention to create a climate-based variety in architecture, with regional characteristics, as opposed to the ubiquitous glass boxes then gaining acceptance.

Interest was growing in non-English-speaking countries also. *Le confort dans l'habitat en pays tropicale* by Dreyfus was published in 1960. Lippsmeier wrote his *Tropenbau* in 1969. In 1973 we published our *Manual of tropical housing* (with Koenigsberger et al.). Today the bioclimatic design approach is well established everywhere, but use of the term itself is most prevalent in Latin countries. The Brazilian architect Marcondes Lima (1995) wrote his PhD thesis on "The development of bioclimatic design".

3 Energy Conservation

The 1973 "energy crisis", triggered by the OPEC oil embargo, brought energy issues centre-stage. The US Office of Emergency Preparedness suggested that there are two approaches to energy conservation and coined the two phrases:

- *Leak plugging:* reducing waste, improving efficiency, without reducing standards
- *Belt tightening:* reducing standards, e.g. providing 500 lux illuminance instead of 1000, heating to 20 °C indoor temperature instead of 23 °C or cooling to 23 °C instead of 20 °C.

Many thought that the latter is not really a "reduction" (or that reduction could mean improvement), but rather a rationalisation of our rather profligate environmental standards. In building equipment "leak plugging", i.e. the improvement of efficiency, became a priority, together with various systems of heat recovery: heat exchangers and heat pump applications. In building terms leak plugging usually meant improved thermal insulation, but it had obvious links with climatic design.

4 Solar-Heated Houses

The utilisation of solar (and other forms of renewable) energy emerged as an important research and development task. The US solar energy research budget jumped from practically nil in 1973 to 400 million dollars by 1978 (still only 4% of the total Department of Energy budget of 10 billion). The International Solar Energy Society, the membership of which consisted primarily of scientists and engineers, at its 1973 Paris congress (supported by UNESCO) debated whether architects could be admitted as members. They decided that any interested person could become a member and the membership more than doubled within a year.

By the end of the 1970s over a thousand solar houses had been documented. Most of these incorporated an "active" solar system, in most cases a scaled-up version of the domestic solar water heater, consisting of flat plate collectors, with a circulating heat transfer fluid (water or air) and some form of thermal storage, mostly water tanks or bins of crushed rock (Fig. 4).

A flood of solar energy publications appeared on the market. The book by Duffie and Beckman, Solar engineering of thermal processes, became the most respected of these, first published in 1980. The new science and technology of solar heating was rapidly developing. The main architectural problem with such active systems was how to integrate the large areas of collectors with the building and how to accommodate the thermal storage (Szokolay 1980).

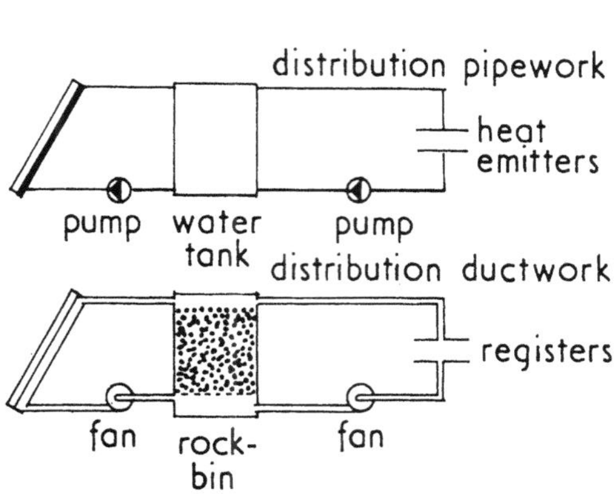

Fig. 4. Active solar systems diagrams

4.1 Passive Systems

The late 1970 saw the emergence of "passive solar systems". Whilst active systems consist of mechanical devices and use some external energy source to drive the heat transfer mechanisms ("parasitic" energy, e.g. that used in pumps, fans or "blowers"), in passive systems the heat transfer takes place by spontaneous processes, without any parasitic energy input. Some authors set the criterion for passive systems as those where the heat transfer takes place by "natural" processes. A more precise definition was offered by Yellot (pers. comm. 1980).

If the coefficient of performance of a solar heating system is CoP = E_s/E_p i.e. the ratio of useful solar energy delivered (E_s) to the parasitic energy (E_p) used to drive the heat transfer processes, then the following system types can be distinguished:

Active systems:	CoP ≤ 20
Hybrid systems:	20 < CoP < 50
Passive systems:	CoP ≥ 50

In a passive solar system the collector, the storage and distribution functions are provided by the building itself (although even passive systems are "allowed" some parasitic energy use, e.g. for adjustments, such as motorised dampers or blinds or movable insulation). The house is the system.

Mazria wrote his *Passive solar energy book* in 1979. He and several other authors produced graphic classifications of passive solar systems, such as that showing the four basic passive systems in Fig. 5.

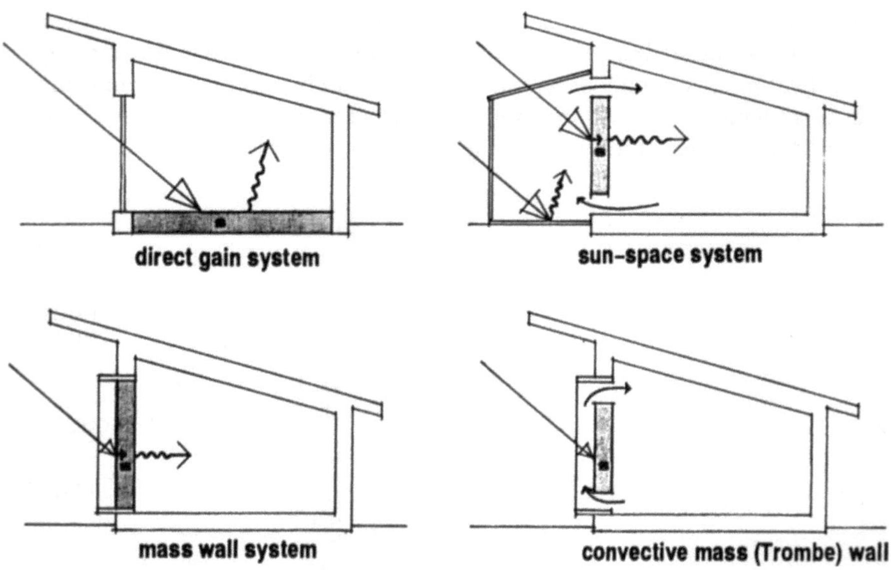

Fig. 5. The four basic passive systems

The **direct gain** system is essentially a large window, oriented towards the equator. Solar radiation entering through this window will be absorbed by the floor or other surfaces (possibly after one or more reflections). The re-radiation emitted by these surfaces will be at much longer wavelengths (≈10 μm), for which the window glass is practically opaque, so the solar heat will be "trapped" in the room. This is indeed the original meaning of the term *greenhouse effect*.

The **sun-space** system may be an attached greenhouse, or some other room with large glass areas, perhaps a glass roof, which acts as a solar heat collector. The warm air from the top of this space is allowed to enter the adjacent room. This warm air flow may be assisted by a fan. Through a low-level opening the cooler room air will be drawn into the sun-space.

The **mass wall** system is basically a heavy masonry wall with a dark surface which will absorb much of the incident solar radiation, protected by a glass skin to reduce the heat loss. The heat is then transmitted through the thickness of the wall end emitted into the room, by convection and low-temperature radiation, with a time delay of several hours (depending on the thickness and material of the wall).

The **convective mass wall** (or Trombe wall) is similar to the above, but it has vent openings to the room near its top and near the floor level. Air in the (approx. 100-mm) gap between the wall surface and the glazing is heated, will rise and enter the room at high level, drawing in cooler air from the room near the floor. This provides warm-air heating to the room during the sunshine period and a delayed heat input through the thickness of the wall. The vents may be fitted with a closing mechanism, which allows the regulation of the relative magnitude of warm-air and the storage-type heating.

In both versions water in containers may be substituted for the masonry mass. This may then be referred to as the **water-wall** system. Whilst these are the basic types, a very large number of permutations and variations are possible; some authors list as many as 24 system types.

A controversy between exponents of active and passive systems went on for some years. The former was the engineers' favoured system, whilst architects were primarily the protagonists of the latter. Passive systems engendered an almost religious zeal; the term "passive" became almost synonymous with "good". What helped the passive systems in gaining acceptance was the fact that passive solar houses need not cost much more than an ordinary house, so they are economically competitive.

5 Design Methods

The heat transfer processes in solar heating systems have been meticulously analysed and algorithms for the simulation of such processes became available by about 1974. However, even with the rapid development of computers such a simulation exercise was a major undertaking, more appropriate to research work than to day-to-day practice of engineering or architectural design.

Prompted by this realisation, a number of simplified design methods have been developed for practical use as design tools. Very large parametric studies were carried out, using sophisticated simulation programs, the dominant parameters were identified and the system performance results correlated with these parameters, usually expressed as dimensionless ratios. These correlation coefficients were then tabulated (or given as graphs), forming the basis of a design tool. For active systems the most widely accepted such design tool is the *f-chart method*, developed by the University of Wisconsin Solar Energy Laboratory (Beckman et al. 1977). For passive systems the *SLR method* (solar-load ratio method), developed by the Los Alamos Scientific Laboratory, became a widely used design tool (Balcomb 1980).

Whilst in active systems there are well-defined and controlled energy flows, the passive system embodied in a building is far more complex, the energy flows more diffuse; one is dealing with numerous ill-defined and low-intensity heat transfers. Consequently, the SLR method has many more limitations than the *f*-chart. It is quite a reliable performance predictor for archetypal passive systems, but the design variables in a real building are so numerous that the "nearest" system for which the tool is available may not be very near to the actual system considered.

It is possible to provide automatic controls for an active system. In passive systems one can only control certain components or parts of the system; the overall performance is the result of the thermal interaction of all parts of the building and it strongly depends on user behaviour. This is the reason why the most sucessful passive solar houses are those designed by architects or engineers for their own use: they adapt their behaviour to bring out the best of the system. As the lady-owner of a passive solar house, full of adjustment mechanisms, crank handles to adjust louvres and ropes to open and close blinds, dampers or vents, said in answer to my question, "...Works well if I sail her properly".

6 Integration

The wheel seems to have turned a full circle: we now realise that the consideration of passive systems in solar houses was an abstraction and only looking at part of the problem. Solar design is a part of climatic design – albeit an important part. Design methods which focus on the solar heating effects are bound to ignore many other (even thermal) factors. On the other hand, every house is a solar house, receiving a solar heat input, although perhaps not making very good use of it.

Not even a well-defined active system can be modelled as a separate (closed) system: it interacts with the thermal system of the building; its performance depends on the building as a whole and on user behaviour. Climatic elements other than solar radiation will have an influence on the thermal behaviour of the building and thus also on the solar system.

The necessity for integration is emphasised for example by Balcomb (1993). Integration is necessary at many levels:

- in active (domestic hot water, heating, cooling and ventilation) systems, between solar and auxiliary parts[2]
- between active and passive systems
- of heating, cooling and daylighting[3]
- of the seasonal behaviour of these systems (e.g. to avoid the passive heating system causing overheating in summer)
- of the complete building and its installations with the users' functions.

It is now agreed that any modelling must also take into account the response of the complete building, including its possible solar system. And, as the building's behaviour is a function of the climate, we are back to climatic design.

The 20-odd years of analytical studies of various active and passive solar systems are now considered as part of a natural progression, their results are not wasted, but are now absorbed into a higher level integrated consideration (dare I say, in a way similar to how Newtonian mechanics having been absorbed into relativity).

7 The CPZ Method

Givoni (1969) used the psychrometric chart as the basis of his building bioclimatic chart. On this he shows the comfort zone and its extensions by various active and passive techniques. However, some of his boundaries are rather arbitrary and not accurately defined. Arens et al. (1981) revised the bioclimatic chart on the basis of the J. B. Pierce laboratory human thermal model. Watson and Labs (1983) further developed Givoni's system.

Szokolay's (1986) CPZ method is superficially similar, but has some basic differences. To define the comfort zone on the psychrometric chart, the Auliciems (1982) expression for thermal neutrality is used, recognising the influence of acclimatisation. The zone is then taken as 4 °C wide (from (T_n – 2 to T_n + 2), plotted on the 50% RH (relative humidity) line and the side boundaries taken as the corresponding sloping ET* (or standard effective tempeature) lines. The humidity limits are set in terms of two horizontal lines of vapour pressure (or moisture content, i.e. absolute humidity). This is usually done for the warmest and coldest months of the year. The comfort zone is not "extended", but the

[2] E.g. a compressor-heat pump can be used for cooling in the summer and (in reverse cycle) for heating in winter; both the exhaust air and waste hot water can be used as sources for such a heat pump, or indeed an unglazed solar absorber panel.

[3] A lack of integration is apparent in a school building where the overdesigned shading device reduces daylighting to such an extent that electric lighting must be used all the time, which substantially increases the air conditioning load in summer; the same device also deprives the building of solar heat input in winter, when it would be desirable.

range of outdoor conditions under which each passive control technique would have the potential of keeping the indoors comfortable are outlined. These are the *control potential zones* (CPZ), giving the name of the method.

There are four basic effects, four basic passive design strategies one can rely on to control the thermal behaviour of a building:

- **Passive solar heating,** e.g. a "direct gain" system: Essentially, large windows with the right (preferably equatorial) orientation, with internal absorbent surfaces and some thermal storage mass, but possibly other passive solar heating systems.
- **Mass effect:** Heavy building envelope elements, which will act as a "thermal flywheel", evening out the diurnal temperature variation. In a hot-dry climate this can be assisted by night ventilation and day-time closure.
- **Air movement effect:** Not only convective cooling, i.e. the removal of internal heat by cooler outside air, but dominantly the physiological cooling effect by enhanced evaporation from the human skin.
- **Evaporative cooling:** Either "passive" (e.g. a pond or a fountain in a courtyard or a spray over the roof), or "hybrid", (e.g. a fan drawing air through wetted fibrous pads: the air delivery is active, but the cooling is passive). The latter may be direct (where the cooled, humidified air is fed into the interior) or indirect. In this system the exhaust air is evaporatively cooled (e.g. by a set of atomiser sprays), and passing through an air-to-air heat exchanger will cool the incoming supply air, without mixing or adding moisture, after which it is discharged (a cutaway view of such a cooler is shown in Fig. 6).

For each of these controls, the outline of the corresponding CPZ is shown in Fig. 7, drawn over the climate representation of three different locations. The climate in each case is depicted by 12 monthly lines, each drawn between two points; the mean maximum temperature with the afternoon humidity and the

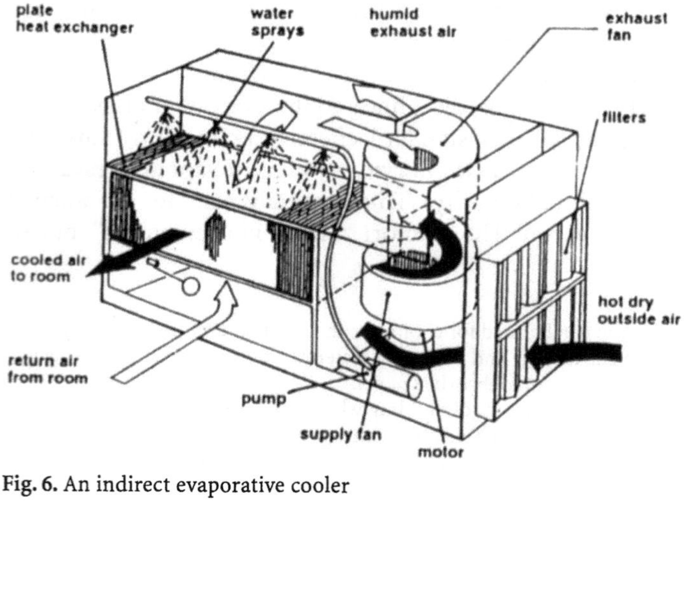

Fig. 6. An indirect evaporative cooler

mean minimum temperature with the morning humidity. The relationship of these lines to the comfort zones indicates the nature of the thermal problem. The control technique with the greatest chance of success is the one, the CPZ of which covers most of the climate lines outside the comfort zone.

While such an analysis can be carried out by manual calculation and plotting, a computer program is available to do this work. It will produce these diagrams, but it also calculates the aggregate length of the 12 monthly climate lines as well as the length below comfort, above comfort and above the humidity limit. The ratios of each of these to the aggregate length give the fraction under-temperature (fuT), the fraction over-temperature (foT) and the fraction, over-humid (foH) respectively. These numbers (displayed at the top left corner of the graphs) give a quick indication of the nature of the climatic problem.

This CPZ method completes and brings up to date the work started by Olgyay, with his bioclimatic chart, in facilitating the examination of the climate/human being/building relationship.

8 Recent Developments

In 1974 the OECD established the International Energy Agency (IEA). Its Committee on Research and Development has a Working Party on Renewable Energy which, in 1989 initiated its Task 13, Advanced Solar Low-Energy Buildings, within their Solar Heating and Cooling Program. Under this Task, 16 houses were designed and built in different climates (in 14 countries, from Japan to Finland), with the aim of reducing the purchased energy consumption to not more than 25% of what comparable present-day houses would use, i.e. to between 30 and 50 kWh/m^2 year. The performance of these houses has been monitored and evaluated and most have been open to the public as demonstration of what is possible. For example, two houses built in the US achieved over 90% reduction in both cooling and heating energy use, resulting in 73–75% reduction of the total energy use (measured as 45–46 kWh/m^2 year). This success was largely due to a very good insulation of the building envelope, with U-values (transmittances) of 0.12–0.2 $W/m^2\,K$, and to the use of an integrated mechanical system for domestic hot water, auxiliary space heating, ventilation and cooling. The thermal mass of one of these houses facilitates night-ventilation cooling in the summer and the storage of daytime solar heat input in the winter.

Some new components were tested as part of these projects, but the main emphasis was on the novel and rational arrangement of known components and design concepts, to minimise energy loss and utilise solar energy (by both active and passive systems) to form an integrated system appropriate to the climate.

Earlier (1986–91), Task 11 of the Solar Heating and Cooling Program examined a series of non-domestic passive solar buildings. The results were published in the form of a "source book" (Hastings et al. 1994) illustrating numerous solutions with technical details and performance data. Besides refi-

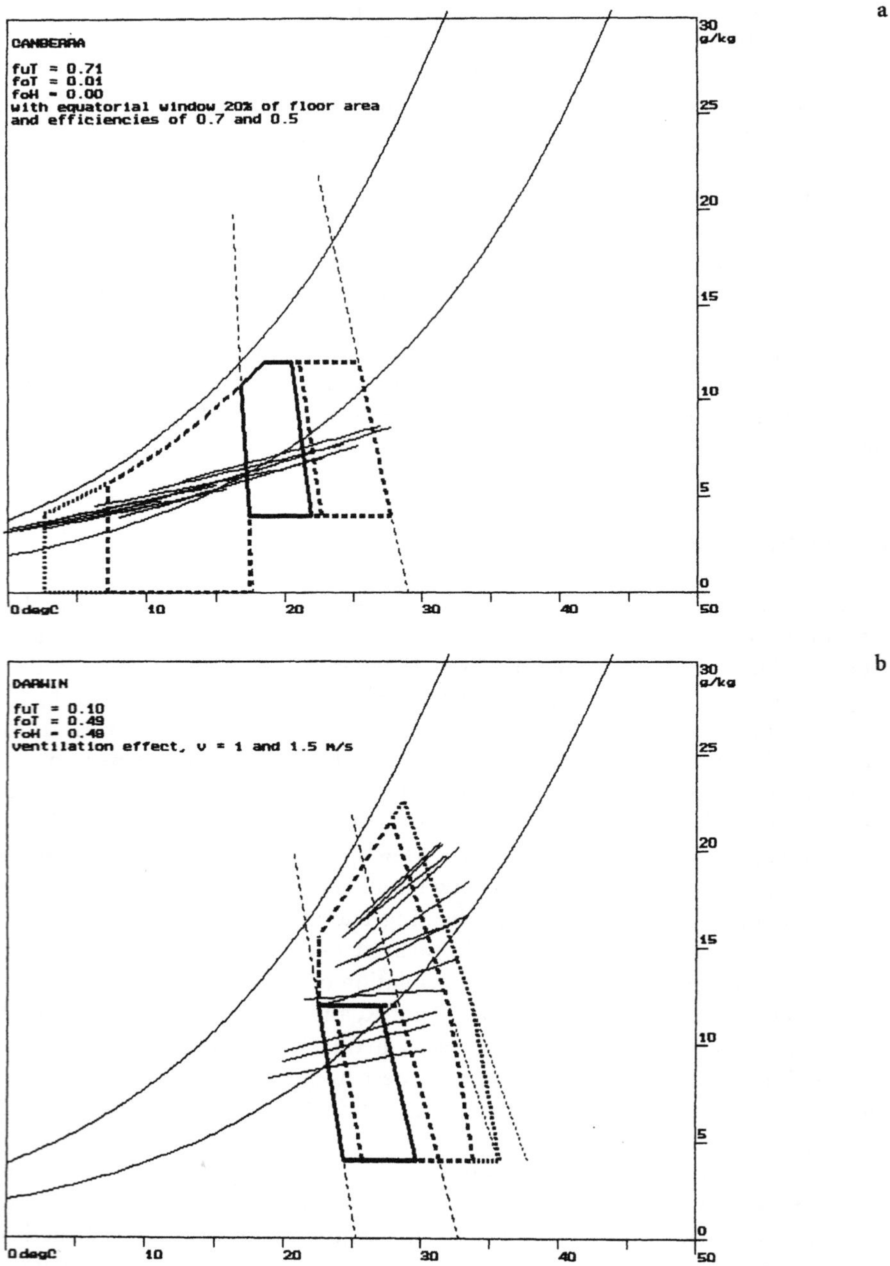

Fig. 7a, b. CPZ diagrams. In each case the comfort zone is outlined for the coldest and hottest month and the 12 monthly climate lines are shown. **a** The potential of passive solar heating in Canberra. **b** The potential of air movement effect in Darwin

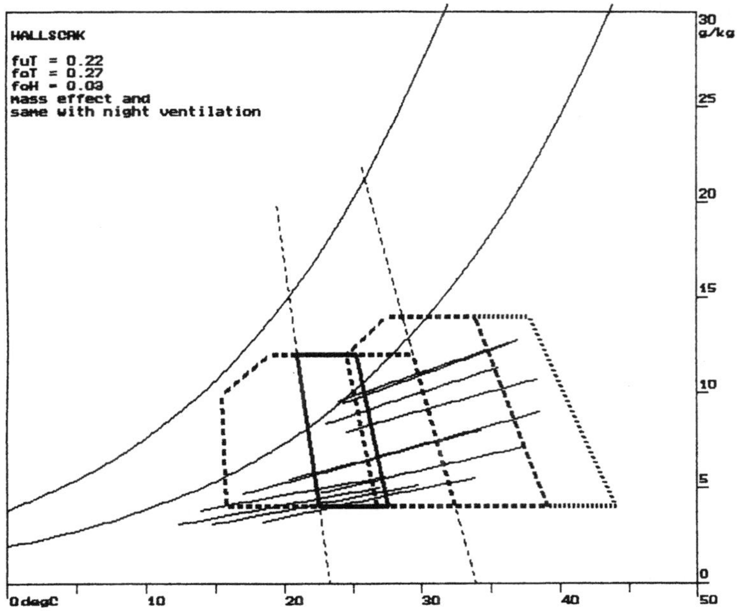

c

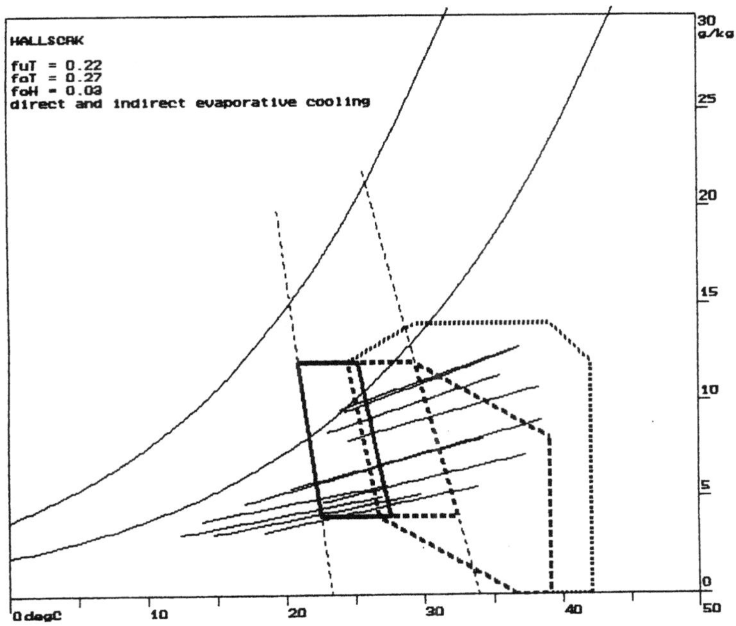

d

Fig. 7c, d. **c** The mass effect (and same with night ventilation) in Halls Creek. **d** Direct and indirect evaporative cooling, the same location as **c**

nement of known passive systems, the book reports on some new techniques, such as

- Transparent insulation used over solid walls (Fig. 8), which thus become solar heat collectors
- Switachable (electrochromic) glazing (Fig. 9)
- Enhanced use of the building fabric for heat storage, e.g. embedded pipes for fluids or air ducts
- New daylighting solutions, e.g. prismatic glazing, light shelves, top lighting and light guides
- Atrium buildings

The European Commission is also very active in this area, primarily in disseminating information to professionals and decision-makers in the form of "resource material". Directorate General XII for Science, Research and Development produced a series of climate-responsive design guides, e.g. the five-volume Design of educational buildings (Yannas 1994). This introduces the principles, gives examples, provides a data-base of documented bioclimatic educational buildings and a bibliography, and also gives working details of innovative components.

Directorate General XVII for Energy, under its THERMIE programme, supports research and development, but it is also concerned with the dissemination of information in the form of reports, books, seminars and symposia, videos, computer diskettes and the Internet.

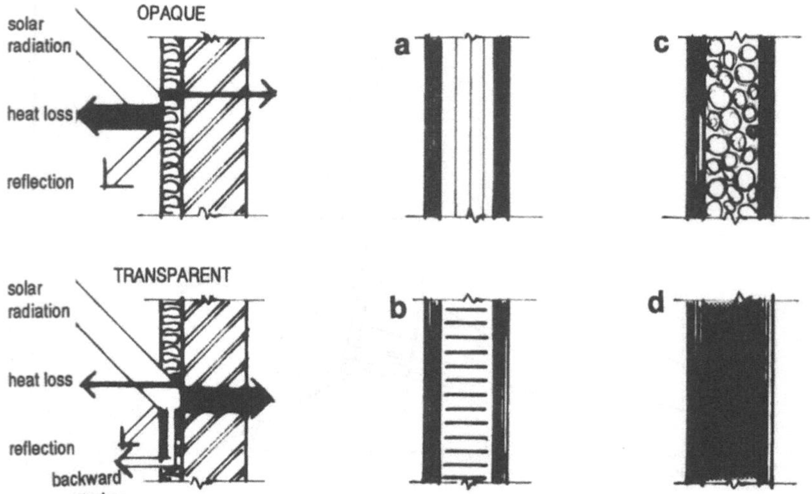

Fig. 8a–d. Transparent insulation used on a massive wall: its effect compared with that of opaque insulation and the four basic types of such insulation (between two rigid transparent sheets): **a** transparent films, parallel with the surface; **b** a honeycomb structure; **c** foam (bubbles, granulated aerogel) or fibres; **d** microporous homogeneous aerogel

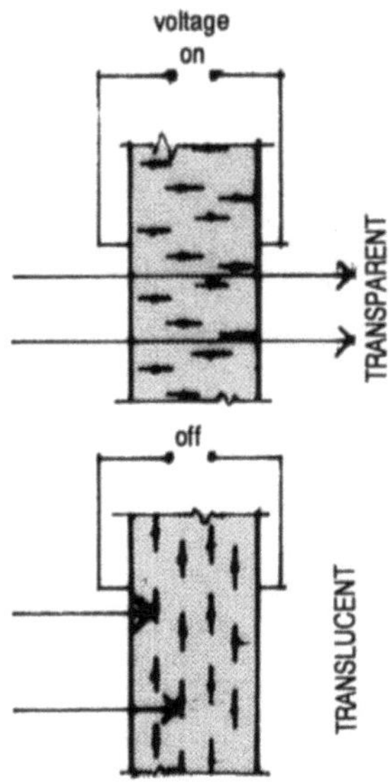

Fig. 9. The principles of switchable glazing (liquid crystal type)

9 Current Issues

Although research and development work is continuing, the greatest effort is directed at the dissemination of information, education of professionals and persuasion of the consumer public, as well as attempting to influence legislators and code-makers to enforce the implementation of proven and developed techniques. The aim is to make bioclimatic design the normal practice.

We tend to think that we know all the technical answers on how to design houses for people to take best advantage of the climatic conditions, both in terms of providing the best possible comfort conditions and in minimising non-renewable energy use. There are, however, several areas where our understanding is incomplete, especially where human factors are involved. Several of these will be examined in the following sections.

9.1 Comfort: Humidity Limit

There is a reasonable agreement on the limits of thermal comfort in terms of temperature. Not so with the limits of acceptable humidity. The best illustration of the changing state of views is to look at the history of the ASHRAE (American Society of Heating, Refrigerating and Air-Conditioning Engineers) 55 thermal comfort standard. Up to the 1966 edition the side boundaries of the comfort zone were vertical DBT (dry bulb temperature) lines (converted to SI units these were 22.9 and 25.3 °C) and the humidity limits were defined in terms of RH, as

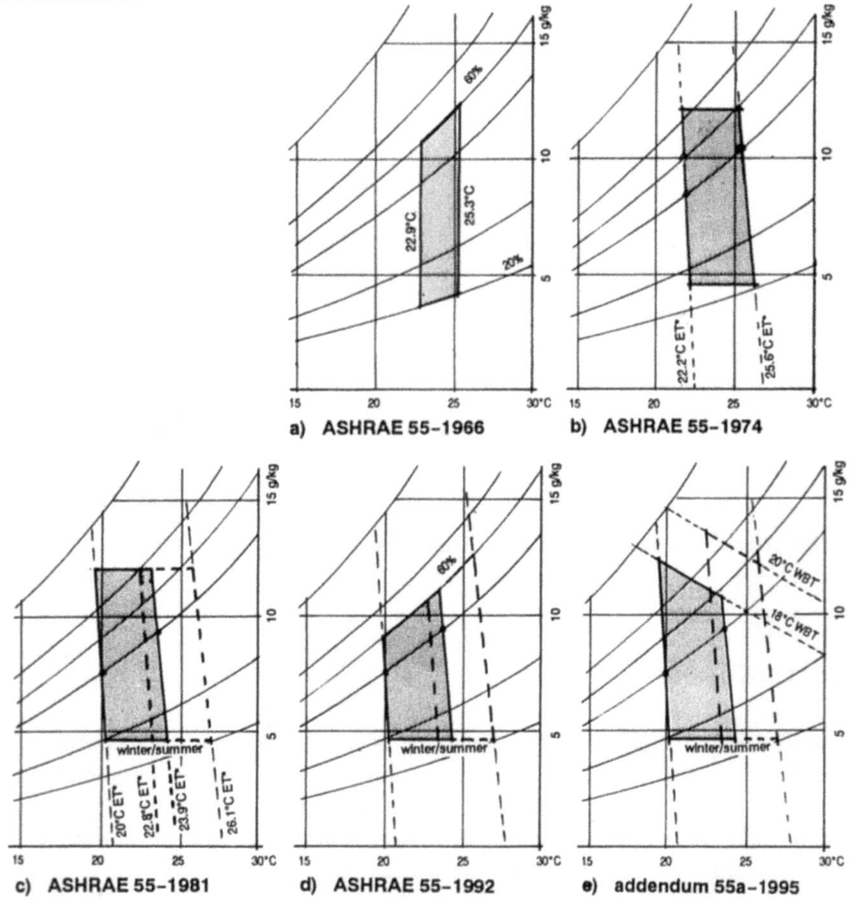

Fig. 10a–e. Changes of the ASHRAE comfort zone definition over the last 30 years

20% and 60% (Fig. 10a). Figure 10b shows the 1974 version, where the ET* lines[4] were substituted as temperature boundaries: 22.2 and 25.6 °C. The humidity limits were defined in terms of dew-point temperatures: 1.7 and 16.7 °C. These correspond to 0.666 and 1.866 kPa vapour pressure and to 4.1 and 11.7 g/kg absolute humidity.

The revision of this standard, confirmed as the 1981 issue, retains the humiditiy limits, but a distinction is made between winter and summer thermal comfort limits: 20–23.9 °C and 22.8–26.1 °C ET* respectively[5], allowing for seasonal differences in clothing (Fig. 10c).

[4] Standard effective temperature lines: these coincide with the respective DBT values at the 50% RH curve.

[5] The horizontal scale in this revision is labelled in terms of OT (operative temperature), which is the same as DBT if surrounding surface temperatures are the same as or close to the air temperature.

Figure 10d shows the 1992 revision, where all temperature limits and the lower humidity limit are retained, but for the upper humidity limit they revert to the 60% RH curve. With respect to this limit there seems to be a complete turn-around in the latest version, the 1995 addendum. Here the upper humidity limits are defined in terms of wet-bulb temperature: 18 °C WBT for winter and 20 °C for summer (Fig 10e).

Apparently ASHRAE recently comissioned a research project (RP-860) at the Lawrence Berkeley Laboratories to investigate thermal comfort at high humidities, and the University of Santa Catarina at Florianopolis (Brazil) is also examining tolerance limits to high humidities.

9.2 Comfort in Solar Houses

It has been shown that termal sensation within buildings depends not only on air temperature, but it is also strongly influenced by the surrounding surface temperatures. This can be measured by the MRT (mean radiant temperature): a solid-angle weighted mean temperature of these surfaces. The concept of environmental temperature was proposed by the UK BRS in the mid-1960s (Loudon 1968), defined as

$$t_{env} = {}^{2}/_{3}\ \mathrm{MRT} + {}^{1}/_{3}\ \mathrm{DBT}.$$

Subsequently (CIBS, 1978) this was found to be valid for people wearing light clothing (e.g. short-sleeved shirts), but for people wearing normal winter indoor clothing (0.8–1 clo) the dry resultant temperature correlates better with their thermal sensation:

$$t_{res} = {}^{1}/_{2}\ \mathrm{MRT} + {}^{1}/_{2}\ \mathrm{DBT}.$$

Szokolay (1985) points out that this has important consequences for houses with passive solar heating systems. In a conventionally heated building (especially with warm air or convective heating systems) the surface temperatures in winter are normally below air temperature. Thus, if the lower comfort limit were (say) 20 °C, it is not the room air that should be kept at this level, but the t_{res}: the room air should be warmer than this to compensate for the cooler surfaces. Conversely, in a passive solar house (especially with direct gain or non-convective mass-wall systems – see Fig. 5) the surface temperatures are normally higher than the air temperature, so the latter can be allowed to drop below the 20 °C. The thermostat of the auxiliary heater should be set accordingly. This, so the argument runs, can quite substantially reduce the (auxiliary) heating requirements and result in energy savings. But is it true? How will the user react? This is still an open question.

9.3 The Mass Effect: Psychological Limits

The mass effect (mentioned in Sect. 7 above) is an important tool of passive thermal control. In the absence of solar and internal heat gains the mean temperatures of the day will be the same inside and outside (Loudon 1968). If necessary, the internal mean can be increased by designing for an increased solar heat input, or it can be reduced by night ventilation. The technique here is to close all windows and doors after sunrise and open them after sunset. The stored heat would thus be dissipated overnight and the indoor temperature would approach the minimum of the outdoor, occurring just before sunrise. It has been observed however, that on a hot day people may keep their doors and windows closed for some time, but almost all will open up the house by late morning or midday, in spite of the fact that the interior is still substantially cooler than the outdoors. One can speculate that in warm climates people prefer open living, or that perhaps psychological factors override the known physical facts. It would be interesting to mount a research project to find out what degree of outdoor overheating would persuade people to close all openings, or how much cooler the interior would need to be to cause an acceptance of closed windows and doors.

9.4 Energy Rating of Houses

Some designers are concerned that many ingrained habits and expectations are preventing the broad adoption of bioclimatic designs. Most people expect the "front" of the house to face the street. The "front" is taken as the side with the largest openings. This "front" has a symbolic, emotive role, it is the public face of the owner, with considerable prestige value and a subject of pride. (This often leads to a rather pretentious appearance.) There is no problem if the street runs east-west and the house faces the equator (i.e. south in the northern and north in the southern hemisphere). However, serious conflict can arise if, for example, this "front" faces west. The owner/client would rarely agree to face the house sideways and show a blank wall to the street. This shifts the responsibility for climatically sound houses from the designer of the individual house to the subdivision designer, to the planner of the street layout.

Local councils, planning departments and building approval authorities have an important role to play, including an educational role. Partly to assist such authorities, partly to increase public awareness, various house energy rating schemes (HERS) have been proposed. In many countries domestic appliances, such as refrigerators, now must have an energy rating, where e.g. 5- stars would indicate the best and 1 star the least energy-efficient unit. There is no compulsion to produce 5-star refrigerators, which tend to be more expensive; one is at liberty to buy the cheaper, 1-star product, but the buyer must be informed.

Similar rating schemes have been devised for houses and are operative in some countries or states. These are in some instances linked to a "trade descriptions act", compelling the vendor to declare the quality of the house offered for sale. The scientific basis of such a rating system is still a hotly debated issue. In some schemes a certain number of points (positive or negative) are awarded for particular features of the house, e.g. equator-facing windows or roof insulation would attract positive points, but leaky construction and west-facing windows negative ones. The sum of the points so earned would then determine the rating. A more realistic rating uses a computer simulation program and is based on the (calculated) annual energy consumption of the house (normalised to unit area, to allow for house size differences), with an agreed standard use pattern and rigidly maintained indoor environmental conditions. Even if (for example) the house has no air conditioning, the program would calculate the energy use as if such a system were installed and used. Both these are purely physical, energy-based ratings, considering the climate/house/energy use relationship. A different approach, also using a computer program, sets the comfort limits month by month, considers the house as "free-running" (i.e. no heating or cooling), and predicts the indoor overheated and underheated degree hours (Kh). This rates the quality of the house, allowing for seasonal variations of human thermal preferences, and leaves it to the occupant how much energy would be used for heating or cooling. This is indeed nearer to a "bioclimatic rating" scheme for houses.

9.5 Tropical Design

It would seem that we have come full circle. Having started with the problems of tropical design we learnt about and more or less solved the problems of building design in temperate and cold climates. The thermal systems of buildings in such climates are now modelled and this computer-based modelling is used with great confidence in design and optimisation exercises. Not so in the wet tropics.

For an essentially closed building the heat flows can be traced with reasonable accuracy. If a building is open (as in most cases in the wet tropics), the air flow is practically unpredictable. We have computational fluid dynamics and we have wind tunnels, but both use arbitrarily chosen wind and terrain conditions. It is a long step from here to depicting the reality.

Wind itself is unpredictable at the building level: we are in the boundary layer. Proximity of other buildings, even vegetation (which tends to grow and overgrow) can significantly change the positive and negative pressure zones around the building, which will determine the air flow through the building spaces. Air flow change may change the temperature, which in turn changes the heat transfer through the building envelope. Hence to use simulation programs developed for European or North American conditions to model warm-humid tropical conditions is highly questionable. Much research and development is necessary in this area, and I am certain that stochastic modelling will be the only way to develop predictive tools.

10 Conclusions

It has been shown that in present day bioclimatic architecture three separate lines of development converge:

- The concern for **climatic design:** the suitability of designs for the given climate, prompted by post-war tropical building activity
- **Energy conservation:** the reduction of energy use in buildings, prompted by the OPEC oil embargo and, later, by concern about greenhouse gas emissions
- **Solar house design:** the utilisation of solar energy in buildings by both active and passive systems.

Modern bioclimatic architecture, the origins of which date from about 1953, combines these influences, but – with its essentially humanist approach – reinstates the human being as the purpose of all these efforts, both in terms of the physical and mental well-being of the individual and at the level of survival of humanity.

References

Arens E et al. (1981) A new bioclimatic chart for environmental design. In: de Oliveira Fernandes, Woods JE & Faist AP (eds) Building energy management. Pergamon Press, Oxford

Aronin JE (1953) Climate and architecture, a Progressive Architecture book. Reinhold, New York

Auliciems A (1982) Psycho-physiological criteria for global thermal zones of building design. J Biometereol. Suppl to vol 26 IX Int Biomet Congr, Osnabrück

Balcomb JD (1980) Passive solar design handbook, vol 2: Design analysis. US DoE, Washington, DC

Balcomb JD (1993) Integrated design. Paper 6 in Proc. Symp on Solar Energy and Buildings. Academy of Athens/PLEA, Athens

Beckman WA, Klein SA, Duffie JA (1977) Solar heating design by the f-chart method. John Wiley, New York

CIBS Guide, A. 1 (1978) Environmental criteria for design. The Chartered Institution of Building Services, London

Dreyfus T (1960) Le confort dans l'habitat en pays torpical. Eyrolles, Paris

Drysdale JW (1952) Designing houses for Australian climates. Bulletin 6. Commonwealth Experimental Building Station, Sydney

Duffie JA, Beckman WA (1980) Solar engineering of thermal processes. John Wiley, New York

Fry M, Drew J (1964) Tropical architecture. Batsford, London

Givoni B (1969) Man, climate and architecture. Applied Science Publ, London

Hastings SR (ed) (1994) Passive solar commercial and institutional buildings. IEA/John Wiley, Chichester

Koenigsberger OH, Ingersoll TG, Mayhew A, Szokolay SV (1973) Manual of tropical housing and building: Climatic design. Longman, London

Lee DHK (1944) Physiological principles in tropical housing. University of Queensland papers vol 1, No 8, Brisbane

Lima A (1995) The development of bioclimatic design. PhD Thesis, The University of Queensland, Brisbane

Lippsmeier G (1969) Tropenbau: building in the tropics. Callwey, Munich

Loudon AG (1968) Summertime temperatures in buildings. BRS current paper 47/68. Building Research Station, Garston, Watford, UK
Mazria E (1979)The passive solar energy book. Rodale Press, Emmaus, Pennsylvania
Olgyay V (1953) Bioclimatic approach to architecture. in BRAB conf report No 5. National Research Council, Washington, DC
Olgyay V (1963) Design with climate. Princeton University Press, Princeton, NJ
Olgyay V, Olgyay A (1957) Solar control and shading devices. Princeton University Press, Princeton, NJ
Phillips RO (1948) Sunshine and shade in Australia. Technical study 23 (also Bulletin 8, 1963). Commonwealth Experimental Building Station, Sydney
Siple PA (1949–1950) Regional climate analyses and design data: the House Beautiful climate control project. In: Bulletin of the American Institute of Architects, September 1949–January 1951, Washington, DC
Szokolay SV (1980) World solar architecture. Architectural Press, London
Szokolay SV (1985) Thermal comfort and passive design. Chap 5. In: Böer KW, Duffie SA (eds) Advances in solar energy, vol 2. Plenum Press, New York
Szokolay SV (1986) Climate analysis based on the psychrometric chart. Int Ambient Ener 7(4):171–182
United States, Office of Emergency Preparedness (1972) The potential for energy conservation, Washington, DC
Watson D, Labs K (1983): Climatic design: energy-efficient building principles and practices. McGraw-Hill, New York
Yannas, S (1994): Design of educational buildings. In: Climate-responsive architectural design portfolios. Architectural Association, London, on behalf of EC, Directorate-General XII for Science, Research and Development

Migration, Recreation and Tourism: Human Responses to Climate Differences

M. FAGENCE[1] and S. KEVAN[2]

1 Introduction

Two readily identifiable types of human movement and travel are responses to the desire to enjoy the circumstances of a congenial and attractive climate and "good" weather conditions: one is seasonal, temporary migration, the other is tourism. This chapter addresses both of these.

There is both similarity and difference in the rationales for these movements. In the early periods of history, extending almost to the twentieth century, seasonal migration was prompted by two forces: firstly, elite social groups travelled to avoid the discomfort of excessive heat (rarely cold) and oppressive urban conditions, especially in tropical and sub-tropical latitudes (Balsdon 1969; Moorehouse 1983); secondly, similar groups travelled to locations where the general circumstances of climate and weather were conducive to congenial working conditions, to recovery from ill health, or, more simply, were free from pollution (Kevan 1993). As society developed, especially from the mid-nineteenth century, an increasing proportion of the population became able to resource travel and visits, which were expected to provide opportunities for recreation, respite or recovery. This more recent travel phenomenon, referred to as "mass tourism", has come to largely dominate temporal and spatial travel patterns in the second half of the twenthieth century. Such mass movement, whilst fitting easily into Maslow's spectrum of human needs, has become dominated by the desire to engage in hedonistic, "sunlust" experiences (Turner and Ash 1975). Even so, the decision to travel (or to migrate for a season) continues to be influenced by an amalgam of physiological, social, psychological and pathological factors (Trump 1967), and is more complicated than the simplistic assumption of the linkage between "good weather" and pleasure-seeking travel.

It is the purpose of this chapter to explore the nexus between climate, weather and the activities of seasonal migration and tourism, thereby revealing more of the depth and breadth of the interconnectedness of biometeorology and tourism planning than is commonly acknowledged by the actions of some policymakers in tourism planning and development.

[1] Department of Geographical Sciences and Planning, The University of Queensland, Brisbane, Queensland 4072, Australia
[2] Department of Geosciences, John Abbott College, Montreal, Quebec, Canada

The first section of the chapter considers the matter of seasonal migratory movements which dominated travel patterns from antiquity, through the Classical and Byzantine periods to the early decades of the ninetennth century. Much of this movement was in response to the search for health cures (Kevan 1993). From this point, the discussion focuses on the mass movements of tourists as more groups in society (mainly, but not exclusively in the northern hemisphere) have become able to take opportunities to explore, to learn, and to seek enjoyment and relaxation in congenial environments (Rybczynski 1991). The discussion on tourism recognises the push and pull factors - the push factors being the circumstances at home which the tourist seeks to avoid, being drawn if not seduced by the expectation of gratification by the pull factors emanating from the destination. Many of the pull factors are climate- and weather-related. A brief section reviews some of the elements of the traveller's decision-making process. This is followed by two extensive sections which will consider, firstly, some of the indicators and indices used to define zones or regions suited to tourism activity, and secondly, some of the measures taken by tourists to achieve suitable levels of adaptation and acclimatization to the region being visited. A concluding section interprets the need for planners and policy-makers with responsibility for tourism and outdoor recreation planning and management to become better informed about the influence of climate and weather on tourism activity, and the impact of those influences on the competitive balance of travel destinations around the world.

2 Seasonal Migration

Of the two types of human response to the allure of regions of good climate, that of temporary seasonal migration has the most extensive history. In a recent review of the literature on migration Kevan (1993) has distinguished between two parallel themes or sources of motivation for the migration: one is the temporary migration of elite social groups to regions which offer more hospitable climate regimes, and the other is the migration in search of health cures. These are manifestations of the search for climate zones commensurate with desired comfort levels (Sargent 1982), and each type of movement is considered briefly here.

2.1 Temporary Migration to More Hospitable Regions

The earliest records of "the seasonal exodus of the rich from the disease-infested squalor of cities to the cleaner environments of their country estates" (Kevan 1993, p. 113) refer to movements in various parts of the world, including Mesopotamia, ancient Egypt, China and India. Medical advice offered to these elites was persuasive, so that, for example, some groups moved in a sequence which was tantamount to enjoying the conditions of a perpetual spring. Commentaries on the periods of the eigth to fifth centuries BC refer to the seasonal migrations of the Persian, Babylonian and Indian ruling elites (e.g. Snorrason 1964). Although there is little early evidence that the Greek leaderships migrated on a seasonal pattern for medical reasons, it has been suggested (Rice 1967) that the Greeks adopted the Persian inclination to travel to more salubrious, cooler residences in the nearby hill areas or on equable Aegean islands such as Kos.

In his review Kevan (1993) reported that "until modern times, no peoples have practised medical tourism with such fervour as did the elite of the Roman Empire" (p. 114), with seasonal migration to take advantage of the curative properties of clean air and fresh stream water and to escape from the potential of fever in the lowland marsh areas. This fervour for temporary migration to more hospitable regions is detectable in later periods with, for example, the elites of the Byzantine Empire, the Moorish Caliphs, and the Mongol leadership relocating to cooler residences. Other examples of seasonal migration occurred in the periods of European colonial influence in southern and south-east Asia (Spender and Thomas 1948, Moorehouse 1983), and in the periods of newly-won independence in the United States with the migration to Rhode Island by southern plantation owners (Sutton 1980).

2.2 Searching for Health Cures

The second type of movement was based largely on the conjecture that certain climate regimes were "healthy"; these were "tours (made by) the ruling rich, in the hope that a change of climate would, if not cure them, at least alleviate their condition" (Kevan 1993, p. 113). For example, Sargent's (1982) description of the "hippocratic heritage" refers to the growing realization, even at the time of the classical periods, that there could be some linkage between climate and various health conditions. Some medical practitioners advocated a "change of air" to relieve particular ailments, while others recommended travel to mineral springs to take advantage of any curative properties such waters had. The visits to the spas were seasonal for two reasons: firstly, the curative properties were related to the humoral theory (upon which much medical practice was based), which "seemed" to be seasonal, and secondly, because travel to the spa regions was often difficult during the winter season.

In addition to the pursuit of clean air to improve health, many medical practitioners advocated various forms of exercise including horse-back riding and sea voyages, providing the activities were undertaken in the most suitable season. Kevan (1993) has reported extensively on the many advocates of such forms of therapy, including the recommendations for preferred locations, times of year, types of travel, and so on. As medical science improved, some physicians became more explicit with their linkage of climate and weather to the state of health of their patients. In the nineteenth century physicians such as Sir James Clark became advocates of the therapeutic advantages of marine environments and sea-bathing, leading eventually to the establishment of resort towns (especially in Britain, France and Germany). Whilst the locations were relatively easy to choose, dispute raged about the length of the suitable season and the various microclimates within and around the resort townships. There is little published evidence that these various matters of contention were ever resolved conclusively. In fact, the conclusion to be drawn from the mass of medical guidebooks and their various locational advocacies and preferences is that the matters were unresolvable at the state of scientific knowledge available at the time (the nineteenth century). Even now there is conjecture about the linkage of health and climate in all but the clearest examples (see later), and the generality of the early discussions has been superseded by more precision and differentiation between arid (desert) cures, alpine (mountain) cures and temperate forest cures, particularly for the treatment of tuberculosis (Kevan 1995).

The circumstance of migratory movements to regions where the climate was more conductive to standards of good health led to the establishment of two industries. Firstly, the seasonal movements became a regular feature of the calendar, leading to the creation of permanaent townships to cope with the flow of patients and visitors. Secondly, that flow of visitors attracted attention to the climatic, locational and landscape qualities of the local environment, eventually giving rise to a conspicuous tourism industry providing "things to see and do" for the patients, their visitors, and eventually other visitors who were not associated in any way with the medical services on offer. One outcome of this extended attractiveness was the expansion of "the season" for which the various forms of accommodation and the facilities were available for use. Although the climatic conditions may have been the original stimulus for a medical service, eventually the attractiveness of the resort region for general (or perhaps specialized) tourism became at least as important (Pomeroy 1957; Jones 1967; Bernard 1978; Rockel 1990).

In addition to the search for locations with good quality air and water, and in association with the expansion of interest in the facilities and amenities of the inland and coastal resorts for various forms of therapy, trends of hedonism in the twentieth century have led to "sunlust" and the fashion to seek fun in the sun (Turner and Ash 1975). There are few early records of reference to the curative power and health-sustaining properties of the sun, and there are indications that the medical fraternity was ambivalent to the curative values. Progressively through the twentieth century the "sun tan" has become a fashion statement of the new age of tourists and recreationists, underpinning many of the decisions on choice of vacation destination. It is in comparatively recent times that some of the dangers of exposure of the body to the sun have become known, leading to the need for measures of precaution and protection (see below).

As societal structures have changed, the focus on climate as a determinant of movement patterns has also changed from a focus on curative properties and hospitable working conditions to one on conditions in which particular pleasure-related activities are possible. This has become an important element in the decision-making process of travellers.

3 Climate as an Influence on the Decision to Travel

A recent examination of the influence of weather and climate on recreation and tourism (Smith 1993) reported that it is "almost self-evident that tourism is dependent on weather and climate ... In some parts of the world it is the climate itself which is the main feature promoting tourism (p. 398). Despite the self-evident nature of the linkages, there have been few systematic investigations into the relationships between weather and climate and tourism.

The published literature of neither field is too helpful: for example, some basic texts (such as Neuberger and Cahir 1969) do not mention the linkage at all, while Oliver's (1973) review of "climate and man's environment" refers to tourism in a section on tertiary industries. The growing literature on the potential impacts of climate change includes some specialist commentaries on the implications for recreation and tourism (see especially Wall 1993), but a recent discussion of the parameters of and focal issues in tourism research (Pearce and Butler 1993) omitted references to climate, to weather, and even to global warming. Yet, clearly, in even the most rudimentary discussions of the push and pull factors in tourism activity, weather and climate are factors of considerable influence on the decision to travel.

It is appropriate to be reminded that not all tourism and travel is recreational or touristic in purpose, and that not all such travel will be impeded by "poor" weather conditions. Most classifications of travel and tourism (for example, Mill and Morrison 1985) identify at least four major categories:

- Travel for recreation and pleasure, and for self-education
- Travel to visit friends and relatives (VFR)
- Travel to conduct business or to attend conferences
- Travel for any one (or more) special purposes (including visits to medical centres to take advantage of special treatment which might be climate-related)

For at least the VFR and business categories the climate, and the weather experienced within that generalised pattern, may be largely irrelevant as a travel determinant. The weather experienced at the destination can only influence the level of enjoyment or success *at the destination;* it may curtail travel at the destination and make movement there generally uncomfortable, or it may limit the opportunities for travel to particular tourism attractions, but it is unlikely to prevent the travel taking place – unless, of course, the outcome of the weather patterns includes snowdrifts, blizzards, floods and storms so that all movement ceases. In the particular case of business travel to conferences, both the volume of visitors and the choice of venue may be influenced significantly by the expected advantages of particular seasonal climate conditions. By contrast, travel for recreation and pleasure will be significantly influenced by the climate and weather circumstances: this nexus has been described as "weather-sensitive tourism" (Smith 1993, p. 398; and see below). In addition, some of the new forms of environmental tourism (sometimes referred to as "ecotourism") will be climate-dependent. Examples of these new forms include travel to experience particular natural phenomena such as bird migrations, animal gatherings at water holes, flowering of shrubs, and so on, all of which will be season-dependent.

Many studies have been conducted to unravel the tourist and traveller decision-making process, especially to identify any commonality in the process, at least amongst the factors which prompt the decision, and those which give that decision "shape". Gray (1970) has differentiated between travel movements which may be described as "wanderlust" (exploratory, almost migratory travel) and those which may described as "sunlust" (almost exclusively hedonistic). Crompton's (1978) small-scale motivational study determined that "a break from routine" was one of the strongest motivational forces, in part supporting the notion of Grinstein (1955) that "lying on the warm sand, being buried in the sand, being practically in the nude are all examples of pleasures which in themselves represent manifestations of partial regression" (p. 179). Similar conclusions have been reached and recorded by Dann (1977), and Leiper (1984), while Graburn (1983) has referred to inversions in tourism – the attraction whilst on holiday of enjoying better weather than is normally experienced at home, and of relaxing rather than being energetic. Some discussions on travel motivation in the tourism literature focus on Maslow's theory of the hierarchy of needs; seeking after good weather conditions fits easily into his structure. A similar good fit occurs with Plog's (1973) psychographic continuum which links motivation, types of experiences and destination types; the pursuit of a satisfactory climate regime for a vacation falls at the midpoint of the continuum. Other commentators, such as Hodgson (1983) and Van Raaij and Francken (1984), refer to the hierarchical nature of decision-making of travellers and the prominence which climate and weather attain.

The allure of good weather conditions is not lost on the transport and travel promotion industries, with charter airlines capitalizing on the preference for seasonal movements in Europe to the Spanish, French and Italian Rivieras from the land-locked and more northerly countries, and with travel agencies producing information which extols the advantages of climates in exotic locations. Making decisions about vacations has been likened to "buying a wish fulfilment", especially if the anticipated weather conditions are amongst the major influences on choice of destination (MINTEL 1991; Perry 1993). As most recreational travel is booked in response to the information presented in the brochures of tour companies, one of the key items is weather information. The vested interest in selling resorts and holiday travel causes travel companies to use photographs, prose and data to convey a particular image of climate and weather conditions. Perry (1993) has noted that for some destinations there is no difference in the photographic representation of conditions despite seasonal changes; his conclusion is that tour companies provide climate information only for those aspects which are favourable to the resort, leading to the possibility that "the weather expectations of the holiday maker frequently differ greatly from the weather experienced on the holiday, with obvious frustration and disappointment" (p. 411).

Some holiday destinations have attempted to create a favourable image by incorporating a climate term in their name – for example the Costa del Sol. Even if the descriptor is accurate, it may camouflage the intensity and duration of inclement weather when that occurs. The dilemma of this degree of climate-dependency in travel promotion is that the traveller books the holiday on the basis of climate, but at the destination what is experienced is weather (Perry 1993, p. 411). Information about climate and weather at tourism destinations which may influence the choice of destination is available from travel companies' brochures, travel magazines (which provide information about resorts on a selective basis), airline in-flight magazines, holiday weather books (e.g. Bernard 1991, 1992; Foord 1973; Pearce and Smith 1984; Rudloff 1981) and television travel shows. This information, whilst it may be accurate, is seldom set in a systematic scientific context, so that inaccurate conclusions may be drawn, not least in respect of the differences between seasons, the generality and consistency of the weather which will be experienced, and the various comfort levels to be experienced. It is possible that the traveller may miss the implications (which may lie hidden in the information) for some precautions needed to ensure adaptation to the climate and weather to be experienced. It is to these more "scientific" issues that our attention is now turned.

4 Indicators of Tourism-Climate Interdependency

Despite the evident close inter-relationship of vacation choice (especially type, destination and duration) and climate and weather conditions, the published literature does not reveal sustained attention to deriving consensus on the measures of interdependency (Baker and Olsson 1992). Some consistency is evident in the assessment of the weather sensitivity of particular tourism and recreation activities, and in the concentration on the measures of comfort levels; but beyond that there is idiosyncracy in purpose, method and study region. By way of example, three approaches are discussed here briefly, both to reveal the differences and similarities, and to expose the potential for co-ordination and integration (which will be discussed in the concluding section of the chapter). In sequence, consideration will be given firstly to Mieczkowski's attempts to develop tourism-climate indices and climate-derived tourism activity zones, then to a number of Canadian attempts to determine "suitable days for particular recreational activity", and finally to an assessment of weather sensitivity for tourism activity.

4.1 Tourism-Climate Indices and Zones

In his brief review of the climate dependency of tourism activity, Mieczkowski (1990) differentiates between macroclimate and microclimate at the regional and local levels, a distinction which was derived in part from his earlier work on the creation of tourism climatic indices (TCI). Whereas there had been numerous attempts to create climate indices from the point of view of permanent residents in particular regions (for example, Davis 1968 for Britain; Gates 1973 for the United States), Mieczkowski (1985) set out to create indices which reflected what was of special interest to travellers and tourists, namely climatic conditions at particular seasons rather than throughout the entire year. His work represents an extension of that of, for example, Heurtier (1968), who investigated the tourist climate of parts of Europe during the summer period, and Crowe (1975), Gates (1975) and Masterton and others (1976), who classified the climate of selected regions in Canada for the purposes of tourism and recreation.

One of the principal difficulties in constructing the indices is unravelling the complex interactions which occur within the phenomenon of "climate". Mieczkowski argued that unless due consideration was given to the extensive range of components of "climate", the real climatic situation would be distorted and the results or outcomes unreliable. The principal and most difficult task, therefore, was to integrate into a single index the separate components relevant to the tourism experience. Whilst it would be axiomatic that the principal components would be temperature, humidity, precipitation, sunshine and wind, it would become apparent that the significance of any one would be less than the combination of them all, and that the most likely determinant for the tourist would be what could be described as "thermal comfort" not least because it would be this circumstance which would determine the duration of the tourist season and the attractiveness of particular destinations. Some of the significant difficulties encountered by Mieczkowski included isolating the components, standardizing the measurements of them, and settling on a measure for optimal conditions.

Mieczkowski considered six factors to be of significance in the generation of measurements of thermal comfort. These were, in two groups:

1. Air temperature (dry-bulb) 2. Relative humidity of ambient air 3. Mean radiant temperature 4. Wind velocity	as environmental factors
5. Thermal resistance of clothing 6. Activity level (influencing the amount of heat produced in the body)	as human factors

Different combinations of the variables would create different levels of thermal sensation; however, for the practical purposes of constructing the indices Mieczkowski assumed that thermal comfort was a condition of thermal neutrality (after Fanger 1970), and that the principal factors determining that condition would be air temperature and relative humidity (with the other factors being "relegated" to lesser significance, or accorded a neutral measurement, or being discarded). Whilst the environmental factors would be amenable to systematic and consistent measurement, it would be likely that the human factors would be distorted by cultural and psychological factors (such as dietary habits, clothing preferences and fashions, activity preferences and expectations of comfort) (Mieczkowski 1985, p. 222). The tendency to derive precise measurements was avoided by Mieczkowski, who preferred to develop comfort "zones" which were based on combinations of factors extending along a continuum of preferred optimal conditions; this followed the "model comfort envelopes" defined by ASHRAE (1974).

Mieczkowski's index draws extensively on his interpretation of the literature of physiology in a new context (i.e. climate). His argument builds progressively towards the construction of measures of optimal thermal comfort, recognizing that the principal determinants would most likely be dry-bulb temperature and relative humidity, and that different outdoor activities and activity levels would induce differences in the levels of comfort and in tolerance of climate conditions. (In simple terms, the comfort levels in ski fields would be different from those of beach areas). A "correction" to the raw data on temperature and humidity was necessary to reflect an assumed period of maximum outdoor activity (noon to 1600 h) and the distortion of the daily index by the incorporation of night-time measurements (when it is assumed little outdoor activity takes place). This correction embraced Hounan's (1967) assessment that diurnal variation in the comfort level was an important determinant of outdoor touristic and recreational behaviour in both cold and warm climate zones.

Although the Mieczkowski index concentrated on air temperature and relative humidity, it did not ignore the other environmental factors. It recognized that precipitation levels and daily occurrence, time, duration and intensity of precipitation could prejudice a tourist's comfort levels. Despite the significance of these variations, in the final construction of the precipitation index the lack of generally available data restricted the index to the incorporation of data only on total amounts of precipitation. Mieczkowski came to the conclusion that, as tourists almost inevitably considered high levels of sunshine to be unequivocally beneficial and certainly desirable at tourism destinations, the levels of sunshine availability should be considered as positive elements in the construction of the total index. Despite some ambivalent evidence from other studies, Mieczkowski considered that it would be important to include wind velocity as a significant determinant of comfort levels, because it could accelerate the transfer of heat (through turbulence and evaporative cooling) or cooling (by removing the overheated, saturated layer of air).

After omitting some of his original variables in the tourist comfort index, principally because of data deficiencies (especially in some developing countries), Mieczkowski settled for a revised combinatorial index of seven variables:

- Maximum daily temperature (°C)
- Mean daily temperature (°C)
- Minimum daily relative humidity (%)
- Daily relative humidity (%)

- Precipitation (mm rain)
- Daily duration of sunshine (h)
- Wind speed (km/h)

After each of these components was allocated a weighting in accordance with its relative importance for tourist well-being, they were incorporated into a formula such that:

TCI = 4 Cid + Cia + 2 R +2 S + W,

where
TCI = tourism climate index;
Cid = daytime comfort index (composed of maximum daily dry-bulb temperature and minimum daily humidtiy);
Cia = daily comfort index (composed of mean daily dry-bulb temperature and mean daily relative humidity);
R = precipitation (mm rain);
S = daily hours of bright sunshine;
W = wind speed (km/h);
and where the total or "best" performance rated 100.
In generalised categories the TCI values may be expressed as shown in Table 1.

Mieczkowski computed data for every month of the year for 453 meteorological stations around the world to generate world-wide maps to illustrate regional patterns of climatic conditions for tourism. Among examples of the monthly variation of TCI, the calculations shown in Table 2 have been made for Toronto (Canada) and Sydney (Australia); the variations according to season, and duration of each category should be noted. His discussion of the outcomes

Table 1. Approximate definition of TCI values

Index value	Description	Examples (January)
80–100	Excellent to ideal	Most of Central America, West Africa, southern Australia
60–79	Good to very good	Florida, Caribbean, eastern Australia
40–59	Marginal to acceptable	Most of continental USA, Mediterranean region, East Asia
20–39	Unfavourable	Most of Europe, Canada, Amazon basin
Less than 19	Impossible	Canadian and Russian tundra, Southeast Asia

Based on Mieczkowski, 1985, pp. 229 ff.

Table 2. Variations in monthly TCI for Toronto and Sydney

City	Jan.	Feb.	Mar.	Apr.	May.	June	July	Aug.	Sept.	Oct.	Nov.	Dec.
Toronto	27	29	38	50	69	86	86	88	76	56	38	29
Sydney	82	76	76	75	65	50	55	61	74	77	79	84

of the research revealed his disquiet with some of the generalisations, some of the judgements which had to be made, the lack of consistent data, and the arbitrariness of the weighting system he adopted. He commented that the "scores" used to generate the maps of climate regions should not be considered as mathematically precise; rather they should be interpreted as reasonable approximations rendering possible a comparative evaluation of climates from a tourists' point of view – a world-wide indication of the potential for "comfortable" tourism.

4.2 Indicators of "Suitable Days"

The generality of the tourism climatic zones derived by Mieczkowski (1985) is indicative only that there is a match between general climate conditions and potential recreational and touristic activity. Refinement of the indices and of the climate zones may be achieved if seasonal differences (Crowe 1975) and daily variations of weather conditions (Gates 1975) are taken into account. This more sharply differentiated measure of recreational and touristic behaviour is achieved with the concept of "suitable days" (Crowe et al. 1973). A synopsis of this concept is considered here. (*Note:* It is surprising that this "refinement" was not included in the later work of Mieczkowski. This may be indicative of the difficulty of achieving consensus, even within the research community in Canada).

The suitability of a region or destination for tourism activity varies according to, for example:

- The nature of the activity
- The weather conditions required to optimise the feasibility of that activity
- The nature of comfort, safety and general level of satisfaction required by the participants

These matters have been reviewed briefly by Crowe (1975) in his commentary on Canadian research on climate and recreational activity. In his review of recreational activity in the Canadian North West Territory he isolated three qualitative measures:

- The principal climate factors (according to season: temperature, wind chill, wind velocity, humidity index, water temperature, reliability of snow cover, visibility, occurrence of precipitation, cloud cover and so on)
- Seasonal differentiation (including "shoulder periods")
- Participant satisfaction levels (in categories loosely defined as "ideal", "marginal" and "sub-marginal")

The differentiation was achieved by setting a series of benchmarks, and then describing comfort percentiles (after Heurtier 1968) for each of the major forms of recreation and touristic activity. An aggregate general index is feasible, but its usefulness is doubtful (in part, because of the diurnal fluctuations referred to in the "correction" to the Mieczkowski index).

Gates (1975) used a similar approach in his assessment of the tourism and outdoor recreation climate of the Canadian Maritime Provinces, while Masterton et al. (1986) further refined the criteria for defining "suitable days" in their study of the Prairie Provinces. In this later study, the refinements included:

- Prescribing the suitability criteria for a timeframe of not less than 5 h in the period 1000–1800 h daily
- Basing data on weather records for nominated stations, converted to percentages of total possible days
- Deriving regional classifications (i.e. zones) based on a predetermined scale of numbers of suitable days
- Determining seasonal length patterns, using such criteria as (a) the period of snow cover (from the first to the last day of a prescribed depth of snow), (b) the period of reliable snow cover, (c) the period of river and lake ice, (d) the period of mean daily temperatures within a prescribed range, (e) the period of reliable temperatures (maximum and minimum duration probability and variance).

These refinements led to the determination of the minimum climatic conditions necessary to sustain participation in particular recreational and touristic activities (Masterton et al. 1976, p. 86). [See also McBoyle and Wall (1992) for a sensitivity analysis on ski areas in the Great Lakes region.]

The derivation of measures of a "suitable day" represents a refinement of the general classification of climatic tourist zones, facilitating a more detailed suitability assessment according to the specific regional (or more likely, resort destination) characteristics. This is a necessary prerequisite to determining the appropriateness of location of particular tourism resorts within a generally suitable climate region, and contributes to the realization of the tourism planning maxim of Baud-Bovy and Lawson (1977) that "even within a tourist region only certain areas are absolutely necessary for tourist development" (p. 181). In some cases, the resort location feasibility assessment may overlook some of the criteria discussed in this section to define "suitable days".

4.3 Indicators of Weather Sensitivity

The more general matter of weather sensitivity of tourism has been considered by Smith (1993), who focused discussion on four indicators: participation rates, comfort and safety levels, satisfaction levels and the profitability of the tourist industry.

Clearly, an important indicator of weather sensitivity is the suitability of weather conditions for tourists and recreationists to participate in their preferred activities. Research in Canada and Australia (e.g. Crowe et al. 1973; Masterton and McNichol 1981; De Freitas 1990; McBoyle and Wall 1992) has examined the circumstances of recreation climatology, indicating the climate parameters for indulging in activities at the beach, in water bodies, in parkland regions for walking, trekking, climbing, and for snow-based activities. There are two parameters of weather sensitivity. The first is the general climate and seasonal conditions which could be expected to sustain the outdoor activity; this parameter dictates the period of the year during which the activity may take plase. The second parameter is the specific conditions of weather which are necessary to facilitate or enable the activity to take place; this parameter is the more refined indicator because it is indicative of, for example, the "precise" conditions (such as duration of adequate snow depth, air temperature range, wind velocity, visibility levels, precipitation incidence) necessary for a particular activity to take place. These refined indicators will determine the recreational choices made, whether to ski, or to climb, or to swim, or to camp, or to seek alternative indoor recreation. Thus, although the general seasonal circumstance (as might be evident from the work of Mieczkowski) might indicate suitability of a region for tourism activity, the actual indulgence in that activity will be determined by the micro-circumstances of conditions at particular sites and on particular days; these days will be the "suitable days" as defined previously, and the indulgence levels will be the participation rates.

The second important factor is the level of safety of indulging in particular activities, and the bodily comfort level during participation. Whilst general climate conditions might be conductive to participation, actual participation will be determined by a number of anthropocentric criteria such as the recognition that the activity will be safe (for example, that the wind velocity will not overturn boats, or sweep climbers from the face of mountains, or disturb campsites). There is little recorded research on this matter, but it will become of increasing significance as the niche market of adventure tourism expands. Some of the research on bodily comfort levels was canvassed in the previous review of Mieczkowski's work. The most cited work includes Terjung (1966, 1968), Giles (1987) and the general message from these sources is that whilst tourists may experience some bodily discomfort (through heat stress or cold), this likelihood may not deter travel to destinations in the climatically less comfortable regions, with Terjung citing as examples the significant visitor levels to the national parks and forests in the most inhospitable regions of Alaska.

A third important factor in tourism is the level of visitor satisfaction with the weather experienced during the vacation. Whilst the weather may not be the most important factor in attracting to a destination, the opportunity to spend time in the "open air" with the opportunity to take "blue sky" photographs is a significant marketing tool. Harrison and Smith (1989) have gathered together papers which provide an overview of the circumstances of weather sensitivity and services in Scotland, and within that collection, Hay (1989) has argued for investment in companion indoor facilities to cope with the demand for activity if the weather should prove to be too inclement for outdoor activity. In contrast to the general advocacy of "good" weather conditions, Smith (1993) has urged a shift in marketing emphasis in Britain towards the acknowledgement that the British weather pattern has variety, has cloud patterns which may even enhance skyscape photography, and which avoids the extremes of temperature and humidity which produce levels of discomfort (p. 401).

In respect of the fourth factor, the weather patterns are important influences on the profitability of the tourism industry, because the use levels of facilities, the maintenance of booking levels, and, in the sub-markets of support services, the viability of equipment manufacturers, suppliers, retailers, transport operators, hire companies, hotels, restaurants, and so on are all affected by weather factors. However, it is not only the incidence of poor weather conditions which may adversely affect the tourism industry; in some cases, exceptionally good weather may lead to greater participation rates, which will increase congestion levels and may heighten the risk of accident and therefore have a flow-on impact on insurers (see later).

Recent research into some aspects of weather sensitivity, and, in particular, to the linkage of that with circumstances of bodily comfort under a number of different levels and types of activity (including tourism), has incorporated refined levels of scientific measurement of, for example, heat and cold stress (Jendritzky 1984, 1991), and body-atmospheric energy exchange (De Freitas 1990). Related research in this area has focused on the generation of seasonality diagrams (Besancenot 1990), the construction of indicators of the relationship of climate conditions to broncho-pulmonary stress (Rivolier et al. 1967), the revelation of the significance of acclimatization to particular weather conditions (Paul 1972), and the creation of bioclimatic diagrams for the information of selected recreation groups (e.g., for backpackers, by Reifsnyder 1980, 1981).

The significance of some of the work considered in this section will become evident laterin the next section, where there will be a discussion of human adaptation to achieve desired comfort levels (acclimatisation) through thorough preparation before the travel or outdoor recreational activity takes place, and the undertaking of appropriate precautions during that activity. To a significant degree, it is the achievement of desirable human comfort levels in tourism activity and outdoor recreation which is the rationale for linking the studies of biometeorology and tourism.

5 Human Adaptation: Acclimatisation, Preparation and Precaution

The desire to maintain comfort levels, or to avoid climate-related discomfort remains a significant determinant of "where to go" and "what to do" in the decision-making process of tourists. However, with advances in medical practice, and in the design and manufacture of appropriate equipment, tourists and travellers are able to withstand conditions which would be almost impossible without those supports. The range of tourism activities and destinations is being extended as tourists seek to challenge their own ingenuity and capacity to withstand levels of discomfort; this is especially so in the new fields of alternative tourism (Smith and Eadington 1992).

Some adaptation to "hostile" climates is achieved by seasonal migrations (similar to that discussed previously), with, for example the movement of Canadians from northern latitudes to the southern states of the US for the duration of the northern winter (Martin 1994), and of Europeans to the warmer zones of Portugal, Spain and Greece – generally the Mediterranean region; these migrations are eventually overwhelmed by the mass market "invasion" during the traditional summer vacation period (BarOn 1975; Hartmann 1986; McEnliff 1992).

Seasonal migration (Butler 1994) is one means of human adaptation. By far the most common, however, are such means as appropriate preparatory and precautionary health measures and the resort to appropriate equipment and clothing. Travel medicine is becoming a significant specialisation of medical practice. It should be remarked here that not all matters of travel health are weather- and climate-related; some of the health isssues are related more to general insanitary conditions and environmental circumstances than to climate. Even so, some of the indigenous matters can be exacerbated where there is a "hostile" climate. There is a considerable literature emerging on travel health, from the general advisories to travellers (such as Dawood 1992), to more specific reports on the inter-relationship of health conditions and travel (for example Page et al. 1994; Hobson and Dietrich 1994; Clift and Page 1996). In addition, the medical journals are hosting reports on studies of travellers' health with attempts to link the health conditions to various localised factors, including climate (for example, Behrens and McAdam 1993; Cossar et al. 1990). A recent study of deaths of Australian travellers overseas (Provic 1995), reported that in a few cases improved preparation for adaptation to the new local conditions might have "saved" lives. It may be that some travellers are prepared to take risks with their health in regions where the climate may be "hostile", and that even the travel advisories from health organisations and now readily available on Internet (Internet references 1 and 2) will not deter travel.

There are both general and specific matters which need to be considered in human adaptation to "hostile" climates:

- The general bodily response to different climate conditions (including adaptation to "hot" climates, or to "cold" climates)
- "Sun Worship", and its consequences (including skin irritations and eye problems)

5.1 Body Responses to Changes of Climate: Acclimatization

The general circumstance of bodily adaptation to different climate conditions has been reviewed briefly in Adam (1992), and more specifically in, for example, Bell (1990, on tropical medicine), Lucas and Gilles (1984, also on tropical medicine), and Ward et al. (1989, on high-altitude medicine). In his brief commentary, Adam draws attention to the ease with which high-speed transport can move people in a short period of time from a climate with which they are in harmony to one with which they may not be. On arrival at the holiday destination the traveller has to adjust to new circumstances of heat transfer (radiation, convection, conduction) and body temperature, and the health of the traveller in the new environment is determined by the extent to which the signals reaching the brain are reacted to intelligently, through precautions and treatment: this is the process of acclimatization, and it may take place at the destination or may be prepared for before leaving home.

It has been claimed that "the body possesses the mechanism necessary to survive in hostile, hot climates, but time is required to evoke them" (Adam 1992, p. 235), and that this adaptive process may take many weeks, depending on the harshness of the climate. It is the medical recommendation that exposure takes place briefly and gradually in a graded program so that the various body systems (especially the circulatory system) learn to adjust to the new environmental conditions, thereby avoiding the trauma associated with discomfort and stress from "instant" adaptation. For the conventional vacation travel, and even more so for the short-duration business travel, such prolonged periods of adaptation are not feasible. Medical advice is clear: prepare for the travel by fitness training, "artificial acclimatization", and if the traveller is neither young nor fit, then the preparation must be prolonged if the stress encountered once the new climate zone is reached is not to be potentially debilitating.

A number of illnesses experienced by travellers can be traced directly to the failure of the human body to cope with the demands of the new climate regime at the destination. For example, undertaking unusual physically stressful exercise, or prolonged exposure to heat stress, or being over-dressed may induce the condition of heat hyperpyrexia (heat stroke, but not necessarily sun stroke). Different forms of heat exhaustion may be encountered, induced by, for example, water deficiency (lack of adequate water intake in hot climates, where the requirements may exceed previous experience), salt deficiency (where the body's reserves of salt have been depleted through excessive sweating and not replaced through a sensible dietary regime) and the inability of the sweat pores to function efficently (anhidrotic heat exhaustion).

Exposure to the very conditions of sunshine and heat which are sought by the hedonistic traveller may lead to the onset of the illnesses mentioned briefly here, and to the common conditions of prickly heat, heat oedema, heat cramps and heat syncope, all of which may be transient until acclimatisation has become complete.

Although the human body has various mechanisms by which it can acclimatise and adjust to hot climates, it has no such mechanisms for adjusting to cold weather regimes. Rather, the adjustment to cold regimes must be by proper diet (to provide sources of energy), clothing (to prevent heat loss) and shelter (to produce an insulation shell) (Lee and Lemons 1949; Adam 1992). As in the case of travel to hot climates, one of the important determinants of comfort at the new destination is the degree of preparation and precaution. The most commonly encountered climate-induced health risks in cold climate zones include:

- Acute accidental hypothermia, as a result of immersion in cold water
- Chronic hypothermia, as a result of a combination of wearing wet clothing and a high wind-chill factor
- Frostbite, as a result of severe chilling of exposed or poorly covered tissue areas (nose, cheeks, chin, ears, hands, feet)
- Chilblains, as a result of alternate exposure to wet and cold conditions and rapid re-warming

The traveller groups most at risk of many of these conditions are the middle-aged groups engaged in exploring, surveying, hill-walking, trekking or climbing. These are the very groups which are emerging as prime tourism market niches in adventure tourism as the enthusiasm for various types of ecotourism capture the imagination of travellers who have previously indulged in only the hedonistic, mass-tourism activities in urban destinations or at beach resorts. This is a matter which could become of increasing concern for joint research between biometerologists and tourism planners.

Preparation is as important for travel to "cold" climates as to the regions of "hot" climates. The nature of the measures to ease the process of acclimatisation includes a preparatory fitness program, an appropriate diet and acquisition of appropriate clothing, footwear and equipment. Whereas preparations and advanced "artificial acclimatisation" are relatively straightforward for travel to the regions of "hot" climates, Adam (1992) has reported that for cold climates "no acclimatization is possible, so it is vital to be adequately prepared ... with suitable clothing and equipment" (p. 249).

5.2 The Hazards of "Sun Workship"

Lowry (1993) and Perry (1993) have discussed the apparent fixation in the tourist industry on marketing tourism destinations with a "4-S" philosophy: sun, sand, sea and sex. Despite the superficial allure of these elements, especially their contribution to the satisfaction of a hedonistic experience, there is increasing evidence that "today's promotional dreams are rapidly turning into tomorrow's nightmares" (Hobson and Dietrich 1994, p. 26). For the purposes of the discussion here, the focus is on the significance of the sun. Hawk (1992) has reviewed briefly the impacts of exposure to the sun, while Bridges (1990) has examined some of the cancer implications, Munnings (1991) has considered the matter specifically from the point of view of sports medicine, and Prawer (1991) and Rudy (1990) have examined the matter from the dermatologist's perspective. Most holiday and travel brochures suggest that exposure to the sun can improve the quality of life and healthiness; however, the medical reaction to such advocacy is to express concern with advice that "if a tan ... is considered a social necessity, it should be acquired gradually, moderately and carefully to minimize damage and its unpleasant consequences" (Hawk 1992, p. 250).

This process of gradual exposure is not a strategy adopted by most tourists, who seem to believe that they have not achieved value for money for the vacation unless they return home with a sun tan. The sun is one of the captivating elements of the tourist's decision-making process, born as much from the allure of it at the destination as from the lack of it at the home of the traveller; this is one of the push-pull factors in the decision process. Overexposure to the sun is linked to a number of skin cancers – basal cell carcinoma, squamous cell carcinoma and malignant melanoma (Prawer 1991). The dangers of exposure are becoming revealed increasingly in the popular media with news broadcasts, popular magazines and daily newspapers including commentary on UV levels (Jenish 1992). The amount of damage through overexposure depends on the time, locality and circumstances of the exposure, UV levels being highest when the sun is high in the sky, at low latitudes near the equator, and where there are reflecting surfaces (such as snow, water and some metallic surfaces) and an almost cloud-free sky; these are just the circumstances extolled in most travel brochures and sought by most tourists. Almost innocently, Mieczkowski's (1985) climatic index map series draws attention to the most desirable destinations for particular types of tourism experience, which coincidentally, include some of the most UV-damaging regions. Experts in the treatment of sun-related health conditions advise that cool winds, haze, light cloud cover, water (when swimming) and the wearing of thin clothing do little to lessen the damaging effects of the sun. It is the penetration of the skin by UV radiation which damages the nuclei and membranes of cells, although the outcome of the sun exposure process may not be immediately apparent. For most sun-worshipping tourists, the immediate outcomes of sun-bathing include sunburn (sometimes extending to blistering), tanning, and the thickening or hardening of the outer skin layers (hyperplasia). Some of these can be treated readily and in the short term, but it is the more enduring damage which is the serious consequence of exposure to the sun, with degenerative change to the skin (ageing), precancers and cancers causing increasing concern (Bridges 1990).

As the commitment to sun indulgence is so pervasive – even if the most sought-after effect is essentially cosmetic and psychological – there is a need for preparation and precaution. Some preparatory tanning is a useful defensive mechanism in that the sensitive skin becomes sensitised to the exposure which will become the focus of indulgence later. The recommendation of progressive tanning by daily increasing the length of exposure is common (Hawk 1992), provided that the application of protective sun-barrier cream is included in the ritual. For particularly vulnerable areas of the body (such as nose and lips), there is a growing "fashion" to apply highly coloured zinc oxide or titanium dioxide-containing creams and pastes. Page et al. (1994) reported that the most common medications taken on vacations in a sample of tourists to Malta were sunblock creams (75.9%). The common advice is "if you feel uncomfortable and weather-beaten" cover up with suitable clothing. Despite the cosmetic attraction, the medical advice is that too much exposure to the sun is harmful, causing pain, swelling, blistering, scarring (of the skin), occasionally fever and in the longer term, a coarsening of the skin, and cancers of the skin. The delayed effects of constant sun exposure cannot be reversed, except, in some cases, by plastic surgery or the use of skin moisturisers.

The range of medical commentary which directly or coincidentally links particular climate zones to a number of health conditions is increasing. Pettit (1992) has referred briefly and generally to the nature of skin problems which may be encountered by travellers, while Mathews et al. (1991), Russel (1990) and Schull (1987) have reported particularly on common medical problems encountered in tropical latitudes, some of which may be attributable directly to exposure to the sun, or to the combinatorial circumstances of tropical climate regimes, or to contact with plants, animals or insects which are indigenous to particular climate regions. There is no shortage of advice to travellers about many of the potential hazards. For example, both the US State Department and the UK Department of Health offer travel advisories, including climate-related information, and some medical agencies (such as the Royal College of Physicians, London) publish special bulletins on travel health issues. In addition, CDC (Centre for Diseases Control, Atlanta, USA) issues travel health advice through published bulletins and its home page on Internet (Internet reference 3), and the Medical College of Wisconsin (International Travellers Clinic) provides a comprehensive preventive health care service on Internet (Internet reference 4).

There is increasing recognition of eye problems which are linked to climate conditions, and expecially to the exposure to strong sunlight (Taylor 1989; Fison and Ffytche 1992). Damage from sunlight may occur to the surface of the eye (such as conjunctiva and cornea), and if extended over a prolonged period the damage may affect the lens and retina. It is reported that cornea damage is most likely to occur where the strong sunlight is coincident with high altitudes and reflective surfaces such as snow or water. One outcome has been the increasing attention which is being given to the design of spectacles to provide adequate filter and barrier to sunlight-related damage. Even general climate conditions which are conducive to parasite and germ habitats pose hazards to eyes, especially in regions of hot climates (Fison and Ffytch 1992, p. 295). Research about to be funded in Australia will investigate the linkage between sun-induced damage to the eyes and some of the virulent forms of cancers which attack the body and its organs.

5.3 Preparations for Travel

The range of advice being sought before travel usually includes (a) climate and weather at the destination, (b) transport, accommodation and other service options, (c) health and medical services and (d) insurance cover and preparatory actions.

Most of these matters have been considered in previous sections in this chapter. One matter which is of increasing significance and merits consideration at this point is the attention being given to travel and holiday insurance

(Fairhurst 1992; Lear 1993; PATA 1994). The matter of vacation insurance cover is composed of two elements; firstly, insurance in respect of a health problem which occurs during the travel or during the stay at the destination – the insurance cover is for the cost of treatment and, if necessary, repatriation; the second element is insurance cover for the nature of the weather at the destination. In the case of the first type of insurance cover, the recommendation is to "practise safe travel, insure yourself" (PATA 1994, P. 1). Most national federations of travel agencies and national governments recommend to all travellers that the taking out of health and travel insurance cover is a priority. The most common form of travel insurance focuses on costs which could be incurred in the receipt of treatment for illness or a medical condition which was not evident before the travel took place. This form of insurance cover usually embraces medical conditions which are attributable to the climate or to specific weather conditions, albeit coincidentally, in the incidence of contracting disease, or of sustaining an injury as a consequence of an accident which might be occasioned by weather conditions (such as unexpected weather pattern changes which result in excessive rainfall, snowfall, dry spells, fog and so on.) The need for health insurance related to travel and tourism has created a large industry with registered medical service providers such as SFA, Asia Emergency Assitance, and American Express's Global Assist across the world. Most customers take out conventional insurance cover, but with the advent of new, more adventurous tourism activities the demand for specialist, even customer-specific insurance cover, is increasing.

Research referred to by the insurance commentator Lear (1993) indicates the prominance of "the weather" in connection with both special events and personal vacations. He reports that the first weather insurance service was begun in Britain in 1920 with the intent of basing insurance underwriting (and therefore the premiums set) on a scientific and assured statistical basis. The insurance industry tends to tailor-make policies for particular events, not least because weather conditions which might impede one form of event may have only an irritant effect on another. The basis of insurance cover prescribes, for example:

- The incidence of inclemency (e.g. the degree of impending rainfall)
- The duration of the cover (usually from before the event in the period during which the possible attendees are deciding whether or not to attend)
- The time of year of the event
- The location of the event

As the range of activities (special events, and free-choice recreation) is extensive, the insurance industry has attempted to distinguish between the various types and the likely impact of different weather patterns. In addition to insurance cover for the event, the insurance industry has developed "abandonment policies" which permit reimbursement if the event does not commence because of adverse weather conditions. Hay (1989) and Smith (1993) have referred to the need for tourism destinations to be aware of the vagaries of local weather regimes, and to consider seriously the development of alternative (i.e. to wet weather) "indoor" facilities. They point out that although for some event organisers the onset of inclement weather may be detrimental, for some business in the tourism region (including alternative entertainment venues, restaurants, local shops) the "bad" weather may prove to be a boon.

In a world-wide context the insurance industry has taken seriously the need to provide "cover" for tourism activity. General brokerage services are available which "introduce" clients to travel insurance providers according to need. For example, Canadians who fly south during the winter to the warmer climate regions of the southern US, can access special "snowbird" insurance cover, even though few of the policies available cover the likelihood if inclement weather (Internet reference 5). Customised world-wide weather insurers provide policies for individuals and businesses who may suffer from financial loss or income interruption because of the whims of nature. Such customised services cover personal disappointment if a vacation has to be cancelled or is ruined because of inclement weather (usually rainstorms or hurricanes) with policy clauses such as:

> "The Company will insure that from... (date) ... that one out of seven days will be dry at the national weather station located at ... (the nearest to the place being visited) ..."

so that, if every day is a rainy day, a full refund of the vacation expenses may be reclaimed (Internet reference 6). Similar provisions can be devised for the organisers of special events, such that if the temperature is extremely high or extremely low, or if the average hourly temperature is not within a specified range, then the organisers can claim against the policy. A British insurance company (Internet reference 7) will carry insurance cover if adverse weather (as defined in the policy) interrupts an event or renders a venue incapacitated, or impedes access by star attractions, or impedes the restoration of facilities after a storm, and so on. The insurance cover is for cancellation, dislocation, disruption and other unforeseen (or usually not budgeted for) items. The insurance company also carries cover for disruption to individual vacation arrangements.

Tourism and travel insurance cover is available (for both the tourist and the events organiser) for weather conditions which are beyond those which can be reasonably expected [given the general climate pattern, as evidenced in, for example, the world-wide gazetteers such as Pearce and Smith (1984) and the other weather guides listed previously]. There is a distinct nexus emerging in this general discussion between the climate regimes as discovered by scientific study, the patterns of travel preferences as exhibited by hedonistic tourists, and the development of the service industries (especially medical, health and insurance) to provide comfort and confidence for the traveller. As the final section of this chapter will discuss, the suggested nexus is capable of systematic linkage in a formal and positive sense to make cross-referencing more credible than may be the case at the present time.

6 Conclusion

This chapter has reviewed a number of matters which provide evidence of the relationship between important issues in climate and weather, in biometerorology and in tourism. In many published sources referenced in the chapter there were comments about the need for research and investigations which transcend the narrow, formal boundaries of particular disciplines (Smith 1993). There is very little published evidence of interdisciplinary research on the linkages between climate and weather and the human activities of tourism and outdoor recreation. Bibliographic research yields few sources which seem to contribute significantly to the understanding of the linkages; rather, as the bibliographic search for this chapter revealed, there are case studies which "show" the linkages, without explaining them, and there are publications which are set firmly within their narrow disciplinary specialisations. This dilemma, commented upon by Mings (1978), has not been resolved significantly in recent years (Smith 1993). The possible exception to this generalization lies in the scope of interdisciplinary research into the likely impacts of global warming (Wall 1993).

As a conclusion to this chapter it is appropriate to revisit what appears to be the dearth of substantive analysis of the nexus between climate conditions and tourism and outdoor recreation, to canvass reasons for the situation, and to identify crucial study areas which could benefit from mutual co-operation in research by climatologists, biometeorologists and tourism planners. The present circumstances should not be condoned by the research community, because many policy initiatives and decisions, many investments in public and private infrastructure and development, and many spatial plans seem to be made on the basis of poor-quality information. The need to incorporate into the feasibility and impact assessments good-quality information on many of the issues raised in this chapter is becoming increasingly important, expecially as the challenge of alternative tourism extends the frontiers of tourism activity.

6.1 Research Difficulties

The spectrum of research which could address the issues of climate-tourism inter-relationships is both extensive and multifaceted; as a consequence, there is a combinatorial problem, with the difficulties endemic to one discipline becoming exacerbated as they are brought together with other disciplines. The principal problems derive from differences in perspective, technique, methodology, data, flexibility and the vulnerability of the disciplines to institutionalisation and inertia. Some of these difficulties, alluded to in previous sections of this chapter, are re-considered briefly here.

A major problem lies in differences in methodology, particularly between the physical and environmental sciences, the medical sciences and the social sciences. Crowe (1975) had to adjust his assessment of inter-relationships to account for qualitative issues, and Mieczkowski (1985) differentiated between environmental and human factors in his climate index research. A second difficulty is the marked degrees of difference in sophistication and confidence which seem to exist between the various science perspectives in research appropriate to the climate-tourism nexus; for example, in his early assessment of biometeorology Trump (1967) referred to the coalition of pathology, physiology, psychology and sociology, which respond differently from the physical sciences to social and technological advances, with corresponding latitudes of interpretation of data and confidence levels in assessments. Perhaps a pervasive dilemma - and a third difficulty - is the assumption that the nexus is self-evident, and therefore scarcely in need of verification; superficial assessments of this kind, which are not peculiar to this study area, are common in the social sciences, and tend to impede or frustrate substantive research and progress towards understanding. A fourth problem lies in the conspicuous inadequacies and inconsistencies in data; this problem has been commented on extensively in research on seasonality in tourism (e.g. BarOn 1975; Snepenger et al. 1990; McEnliff 1992), and is evident in the simplification forced on Mieczkowski (1985) and the tentativeness with which the Canadian and French researchers of outdoor recreation patterns relate them to climate patterns and weather conditions. One reason for the tentativeness lies behind the fifth problem area; as society develops, and particularly as technological advances become available to tourists and recreationists, the zonal comfort prescriptions which underpin some climate assessments are skewed if there are improvements to travel and recreation clothing, medication, building or facility design. A related, sixth problem, is derived from the willingness of travellers to experiment with their own comfort and safety levels; "the odyssean spirit of wanderlust seems to be overcoming the climatic barriers" (Singh 1975, p. 42, cited in Pearce 1982, p. 153), a matter which is being exacerbated as new tourism areas and activities are encountered, pushing back the frontiers of Maslow's and Plog's spectra of human motivations and satisfaction levels. A seventh difficulty has been referenced by Butler (1994) in his recent examination of seasonality in which he has referred to the institutionalisation of vacation-taking, and the apparent predilection towards inertia of most tourists; these two factors impede rational attempts to extend the period during which comfort levels can be sustained to the satisfaction of tourists (Netherlands 1991).

Thus, some difficulties encountered in unravelling the nexus, and in preparing strategies to cope with tourism and outdoor recreation demand, are derived from the nature and characteristics of the disciplines involved, some are a consequence of the milieu in which these activities take place, some are derived from scientific deficiency (especially in data), and some are the result of the disciplinary puritanism which frustrates co-operation.

6.2 Research Opportunities

It is the disciplinary puritanism, somewhat evident in the exclusiveness of the published information on climate, weather, biometeorology and tourism, which creates one of the most significant impediments to clarification and improvement to understanding. There is some evidence of "re-inventing the wheel". To avoid wasteful duplication of effort and resources, one of the principal opportunities lies in the pursuit of mutual co-operation, facilitated by cross-referencing of research and the achievement of publication beyond the strict disciplinary boundaries. Much effort of climate-tourism research seems to be expended on the global warming focus; there are more immediate concerns which need to be addressed.

From the discussions in this chapter there would seem to be a need to:

- Revisit the tourism and outdoor recreation studies of the 1970s, to conduct post-audits of the assessments and zonal structures created at that time, as a step towards any necessary revision
- Revisit the matter of thermal comfort levels, especially to evaluate the impact of technological advances in equipment, clothing, transport, facilities and so on
- Cross-reference the scientifically derived thermal comfort levels with participation rates, behaviour and attitudes, especially to determine how participants adjust to the extremities of the thermal comfort levels in the new territories of tourism (in adventure and ecotourism)
- Create seasonal weather profiles for tourism destinations, expanding the scope of the factors considered by, for example, Mieczkowski (1985)
- Re-examine, and then monitor, seasonality issues such as institutionalization and inertia
- Examine the impact on vacation-taking and tourism facility provision of the availability of insurance services
- Examine the impacts which medical advice and preparatory and precautionary measures have on traveller decision-making
- Incorporate in climate-tourism zonal patterns the factors of location, accessibility, costs, challenges and so on
- Determine the significance of weather and climate to the Maslow and Plog range of motivations and choice
- Assess the economic implications of changes to climate patterns and weather conditions for public infrastructure, commercial facilities, transport facilities and services and emergency services
- Determine the relationship of vacation-taking to the seasonality of bio-physical resource renewal

In his recent reappraisal of the influence of weather and climate on recreation and tourism, Smith (1993) observed that the reason for the dearth of investigations into the relationships is "that meteorologisits and leisure specialists rarely communicate with each other" (p. 398), in part because, as far as tourism is concerned, the meteorological conditions comprise only one set of factors which influence the decision to travel, to recreate and so on. It is necessary to construct decision models which clarify the impact which these conditions have on travellers' decisions. In tourism, perceptions are important – the conditions expected at the resort destination, the anticipated hedonistic experiences and outcomes (including the sun tan) and all the rest. However, the contribution of climate and weather to satisfaction levels is unclear.

It has become evident, however, from the bibliographic research on which this chapter is based and from the recent reassessment of the subject by some commentators, that many of the published sources on the weather sensitivity of tourism are two decades old; there is a clear need for the matter to be revisited "with a vengeance", not least because there is lost ground to recover and new technologies and societal conditions to infuse into the assessments. Butler's (1994) criticism of research on seasonality is that it "is a subject which is often discussed ... but rarely analysed. One result of this is that there has been little development of theory or concepts pertaining to the topic ..." (p. 328), and much of the discussion is based on supposition and assumption rather than empirical or "pure" research. Interpreted in a broader context, it is appropriate to seize the opportunity set by interest in global warming to undertake *substantive* co-operative research on the inter-relationship of climate, weather, tourism and outdoor recreation. The discussion of issues in this chapter has been designed to contribute to that process.

Acknowledgements. An original version of this chapter was prepared by Simon Kevan. After his untimely death the tourism sections were revised substantially by Michael Fagence, and the entire chapter was rewritten by him with the benefit of comments from Dr. Geoff McBoyle (University of Waterloo) and Dr. Stephen Craig-Smith (University of Queensland)

References

Adam JM (1992) Effects of climatic extremes. In: Dawood R (ed) Travellers' health, 3rd edn. Oxford University Press, New York, pp 232–249
ASHRAE (1974) Thermal environmental conditions for human occupancy. American Society of Heating, Refrigerating and Air Conditioning Engineers, New York
Baker WM, Olsson LE (1992) Tourism: a climate-sensitive industry. Ind Environ 15:9–16
Balsdon JP (1969) Life and leisure in ancient Rome. Bodley Head, London
BarOn RV (1975) Seasonality in tourism. Economic Intelligence Unit, London
Baud-Bovy M, Lawson F (1977) Tourism and recreation development. Architectural Press, London
Behrens RH, McAdam KP (1993) Travel medicine. Br Med Bull 49:2
Bell DR (1990) Lecture notes on tropical medicine. Blackwell, London
Bernard HW (1991) The travellers almanac – North America. Riverside Publishing, New York
Bernard HW (1992) The travellers almanac – Europe. Riverside Publishing, New York
Bernard PP (1978) Rush to the Alps: the evolution of vacationing in Switzerland. East Eur Monogr 37:228
Besancenot J-P (1990) Climat et tourisme. Masson, Paris
Bridges B (1990) Sunlight, DNA damage and skin cancer: a new perspective. Jpn J Cancer Res 81:105–107

Butler RW (1994) Seasonality in tourism: issues and problems. In: Seaton AV (ed) Tourism: the state of the art. John Wiley, Chichester, pp 332–339
Clift S, Page SJ (eds) 1996) Health and the international tourist. Routledge, London
Cossar JH, Reid D, Falcon RJ et al. (1990) A cumulative review of studies of travellers, their experience of illness and the implications of their findings. J Infect 21:27–42
Crompton JL (1978) Motivations for pleasure vacation. Ann Tourism Res 9(4):408–424
Crowe RB (1975) Recreation, tourism, climate – a Canadian perspective. Weather 30:248–254
Crowe RB, McKay GA, Baker WM (1973) The tourist and outdoor recreation climate of Ontario. (Publications in Applied Meteorology REC-1-73), Atmospheric Environment Service, Environment Canada, Toronto
Dann GM (1977) Anomie, ego-enhancement and tourism. Ann Tourism Res 4(4):184–194
Davis NE (1968) An optimum summer weather index. Weather 23:305–317
Dawood R (ed) (1992) Travellers' health, 3rd edn. Oxford University Press, New York
De Freitas CR (1990) Recreation climate assessment. Int J Climatol 10:89–103
Fairhurst R (1992) Health insurance for international travel. In: Dawood R (ed) Travellers' health, 3rd edn. Oxford University Press, New York, pp 371–375
Fanger PO (1970) Thermal comfort. Danish Technical Press, Copenhagen
Fison P, Ffytche T (1992) Eye problems in travellers. In: Dawood R (ed) Travellers' health, 3rd edn. Oxford University Press, New York, pp 292–299
Foord HV (1973) Holiday weather. William Kimber, London
Gates AD (1975) The tourism and outdoor recreation climate of the Maritime Provinces. Problems in Applied Meteorology, REC-3-73, Atmospheric Environment Service, Environment Canada, Toronto
Gates M (1973) Man and his environment: climate. Harper and Row, New York
Giles BD, Balafoutis C, Arseni-Papdimitriou A (1987) A study of climate stress and physiological regions in Greece. J Climatol 7:303–312
Graburn NH (1983) The anthropology of tourism. Ann Tourism Res 10(1):9–33
Gray HP (1970) International travel – international trade. Heath, Lexington
Grinstein A (1955) Vacations: a psycho-analytic study. In J Psycho-Anal 36(3):177–185
Harrison SJ, Smith K (eds) (1989) Weather sensitivity and services in Scotland. Scottish Academic Press, Edinburgh
Hartmann R (1986) Tourism, seasonality and social change. Leisure Stud 51:25–33
Hawk JLM (1992) Sun and the traveller. In: Dawood (ed) Travellers' health, 3rd edn. Oxford University Press, New York, pp 249–255
Hay B (1989) Tourism and the Scottish weather. In: Harrison SJ, Smith K (eds) Weather sensitivity and services in Scotland. Scottish Academic Press, Edinburgh
Heurtier R (1968) Essaie de climatologie touristique. Meteorologie 7:71–107, and 8:519–566
Hobson JSP, Dietrich U (1994) Tourism, health and quality of life. J Travel Tourism Market 34:21–38
Hodgson P (1983) Research into the complex nature of the holiday choice process. Proceedings, Seminar on the Importance of Research in the Tourist Industry. Helsinki, 8–11 June, 1983. ESOMAR, Amsterdam, pp 17–35
Hounan CE (1967) Meteorological factors affecting physical comfort. Int J Biometeorol 11:151–162
Internet references:
1. http://www.cdc.gov/cdc.htm
2. http://nearnet.gnn.com/gnn/wic/wics/trav.travhealth.html
3. http://www.cdc.gov/cdc.htm
4. http://www.intmed.mcs.edu/travel.html
5. http://vaxxine.com/innovative/snowbird.html
6. http://weatherins.inter.net/weatherins
7. http://dspace.dial.pipex.com
Jendritzky G (1984) Bioclimatic maps of heat and cold stress of man. In: Kirchofer W, Ohmura A, Warner H (eds) Appl Climatol 14:101–102
Jendritzky G (1990) Methodik zur raumbezogenen Bewertung der thermischen Komponente im Bioklima des Menschen. Beitr Akad Raumforsch Landspl 114

Jendritzky G (1991) Selcted questions of topical interest in human bioclimatology. Int J Biometeorol 35:139–150
Jenish D (1992) Summer cover-up: fear of skin cancer drives people off the beach and out of the sun. Maclean's 105:34–38
Jones BM (1967) Health-seekers in the southwest 1817–1900. University of Oklahoma Press, Norman
Kevan SM (1993) Quests for cures: a history of tourism for climate and health. Int J Biometeorol 37:113–124
Kevan SM (1995) A change of air (unpublished paper)
Lear J (1993) Weather insurance. Weather 48:108–110
Lee D, Lemons H (1949) Clothing for global man. Geogr Rev 39:181–213
Leiper N (1984) Tourism and leisure. Proc 12th New Zealand Conf, New Zealand Geographical Society, Christchurch, pp 249–253
Lowry L (1993) Sun, sand, sea and sex: a look at tourism advertising through the decoding and intepretation of four typical tourism advertisements. In: Chon K (ed) Proceedings of Research and Academic Papers, vol 5. University of Las Vegas, Nevada, pp 183–204
Lucas AO, Gilles HM (1984) A short textbook of preventive medicine for the tropics. Hodder and Stoughton, London
Martin R (1994) Canadians' favourite sunsport: Florida. Travel-log (Statistics Canada) 13:1–8
Masterton JM, McNichol DW (1981) A recreation climatology of the National Capital Region. (Climatological Studies No 34), Atmospheric Environment Service, Environment Canada, Toronto
Masterton JM, Crowe RB, Baker WM (1976) The tourism and outdoor recreation climate of the Prairie Provinces. (Publications in Applied Meteorology, REC-1-75) Atmospheric Enivronment Service, Environment Canada, Toronto
Mathews DS, Pust RE, Cordes DH (1991) Prevention and treatment of travel-related illness. Am Fam Phys 44:1343–1359
McBoyle G, Wall G (1992) Great Lakes skiing and climate change. In: Gill A, Hartmann R (eds) Mountain resort development. Proceedings of conference, Vail, Colorado (April 1991), Centre for Tourism Policy and Research, Simon Fraser University, Burnaby, BC, Canada, pp 82–92
McEnliff J (1992) Seasonality of tourism demand in the European Community. Economic Intelligence Unit, London
Mieczkowski Z (1985) The tourism climatic index: a method of evaluating world climates for tourism. Can Geogr 29:220–233
Mieczkowski Z (1990) World trends in tourism and recreation. Peter Lang, New York
Mill R, Morrison AM (1985) The tourism system. Prentice Hall, Englewood Cliffs, NJ
Mings R (1978) Climate and tourism development: an annotated bibliography. Climatological Publications, Bibliography Series 4. State Climatologist for Arizona, Tempe
MINTEL (1991) MINTEL special report holidays. MINTEL, London
Moorehouse G (1991) India Britannica. Harvill, London
Munnings F (1991) Sun safety: shedding light on the risks of exposure. Phys Sports Med 19:100–107
Netherlands (1991) Improving seasonal spread of tourism. Netherlands Ministrie van Economische Zak, Rotterdam
Neuberger HH, Cahir J (1969) Principles of climatology. Holt, Rinehart and Winston, New York
Oliver J (1973) Climate and man's environment. John Wiley, New York
Page SJ, Clift S, Clark N (1994) Tourist health: the precautions, behaviour and health problems of British tourists in Malta. In: Seaton AV (ed) Tourism: the state of the art. John Wiley, Chichester, pp 799–817
PATA (1994) Tales of the dark side. PATA (Pacific Asia Travel Assoc.) Travel News, November 1994, pp 9–12
Paul AH (1972) Weather and the daily use of outdoor recreation areas in Canada. In: Taylor JA (ed) Weather forecasting for agriculture and industry. David and Charles, Newton Abbott, pp 132–146

Pearce DG (1982) Tourist development, Longman, Harlow
Pearce DG, Butler RW (1993) Tourism research: critiques and challenges. Routledge, London
Pearce EA, Smith CG (1984) The world weather guide. Hutchinson, London
Perry A (1993) Climate and weather information for the package holiday-maker. Weather 48:410–414
Pettit JHS (1992) Skin problems in travellers. In: In: Dawood (ed) Travellers' health, 3rd edn. Oxford University Press, New York, pp 278–285
Plog S (1973) Why destination areas rise and fall in popularity. Cornell HRA Quarterly, November:13–16
Pomeroy E (1957) In search of the Golden West: the tourist in Western America. Knopf, New York
Prawer SE (1991) Sun-related skin diseases. Postgrad Med 89(8):51–66
Provic P (1995) Deaths of Australian travellers overseas. Med J Aust 163:27–30
Reifsnyder WE (1980) Weathering the wilderness. Sierra Club, San Francisco
Reifsnyder WE (1981) A climatic analysis for backcountry recreation. In: Overdieck B et al. (eds) Biometeorology 8, Proc 9th Int Biometeorology Congr, pp 87–99
Rice TT (1967) Everyday life in Byzantium. Dorset, New York
Rivolier J, Campons R, Lemee P, Wolfromm R (1967) Indices biométéorologiques sur le traitement de l'asthme. Presse Therm Clim 101:137–141
Rockel I (1990) Taking the waters: early spas in New Zealand. Government Printing Office, Wellington NZ
Rudloff W (1981) World climates. Wissenschaftliche Verlagsgesellschaft, Stuttgart
Rudy SJ (1990) Skin cancer incidence and risk. Dermatol Nursing 2:73–78
Russell M (1990) Before departure: advising tropical travellers. Can Fam Phys 36:1297–1301
Rybuczynski W (1991) Waiting for the weekend. Viking, New York
Sargent F (1982) Hippocratic heritage: a history of ideas about weather and health. Pergamon Press, New York
Schull CR (1987) Common medical problems in the tropics. MacMillan, London
Singh TV (1975) Tourism and tourist industry. New Heights, Delhi
Smith K (1993) The influence of weather and climate on recreation and tourism. Weather 48:398–404
Smith V, Eadington WR (eds) (1992) Tourism alternatives. John Wiley, Chichester
Snepenger D, Housere B, Snepenger M (1990) Seasonality of demand. Ann Tourism Res 17:628–630
Snorrason E (1964) Early history of medical climatology. In: Licht S (ed) Medical climatology. Elizabeth Licht, New Haven, pp 594–612
Spencer JE, Thomas WL (1948) The hill stations and summer resorts of the Orient. Geogr Rev 38:637–651
Sutton H (1980) Travellers: the American tourist from stagecoach to space shuttle. Morow, New York
Taylor HR (1989) The biological effects of UV-B on the eye. Photochem Photobiol 50(4):489–492
Terjung WH (1966) Physiologic climates of the coterminous United States: a bioclimatic classification based on man. Ann Assoc Am Geogr 56(1):141–179
Terjung WH (1968) Some thoughts on recreation geography in Alaska from a physio-climatic viewpoint. Calif Georgr 9:27–39
Trump SW (1967) Biometeorology. In: Fairbridge RW (ed) Encyclopedia of atmospheric sciences. Reinhold, New York
Turner J, Ash J (1975) The golden hordes: international tourism and the pleasure periphery. Constable, London
Van Raiij WF, Francken DA (1984) Vacation decisions, activities and satisfactions. Ann Tourism Res 11(1):101–112
Wall G (1993) Tourism in a warmer world. In: Glyptis S (ed) Leisure and the environment. Bellhaven Press, London, pp 293–306
Ward MP, Milledge JS, West JB (1989) High altitude medicine and physiology. Chapman and Hall, London

Climate and Human Mortality: Relationships and Mitigating Measures

L. S. KALKSTEIN[1]

1 Introduction

The impact of climate on human mortality has become a topic of increasing discussion in light of numerous deaths due to heat during the summer of 1995, and the possible impact of a global warming on such mortality (Haines et al. 1993; McMichael 1993; Kalkstein 1995). However, there are numerous uncertainties involving climate/mortality relationships. For example, what is the magnitude of winter weather-related mortality as compared to summer? If the globe warms, will decreases in cold-related mortality compensate for any summer increases? How do confounding social factors, such as increasing air conditioning use, impact heat-related mortality? Are developing countries more affected than developed? Can we acclimatize to global warming? These questions and others will be discussed within this chapter, which will concentrate on how mortality rates are affected by the weather.

Most research on heat/mortality relationships indicates that high temperature can have a dramatic effect on mortality, and other factors such as humidity and wind are somewhat less significant (Ellis and Nelson 1978; Kalkstein 1991b). In fact, the notion of a "temperature threshold", which represents the temperature beyond which mortality rises significantly, is quite common in certain studies, and in many US cities it is apparent that elevated mortality occurs only during the 10–15% of all summer days with temperatures above the threshold (WMO/WMO/UNEP 1996). The height of the threshold temperature appears to be related to its frequency across these cities. For example, the temperature exceeds 36 °C in St. Louis with approximately the same frequency that it exceeds 32 °C in Detroit, values which represent threshold temperatures for these cities. This strongly suggests that the notion of a "heat wave" is relative on an interregional scale, and is dependent upon the frequency of the threshold temperature.

A most interesting finding related to heat-induced mortality which potentially could have a profound influence on such deaths in a warmer world is the large degree of variation in response on an inter-regional level. Recent studies demonstrated that many cities in temperate regions, where hot weather is severe but irregular in occurrence, show a sharp rise in total mortality during

[1] Center for Climatic Research, Department of Geography, University of Delaware, Newark, Delaware 19716, USA

unusually hot weather conditions (Smith and Tirpak 1989; Tan 1994; IPCC 1996). In some cases, daily mortality can be more than double baseline levels when weather is oppressive. Conversely, cities in more tropical locales seem less affected by hot weather. One possible explanation for this interregional disparity may involve the variance in summer temperatures across regions. In the northern and midwestern cities, the very hot days or episodes are imbedded within periods of cooler weather. Thus, the physiological and behavioral "shock value" of a very high temperature is considerable. In tropical cities, the hottest periods are less unique, as they do not vary much from the mean. This seems to play a role in diminishing the impact of a very hot episode on human mortality.

The variety of factors affecting weather-related mortality has led to a number of recent studies. Most are concerned with heat-related issues, but cold-related mortality has been attracting increasing interest. What follows is a more detailed evaluation of some recent research, which should demonstrate the complexity of interacting factors which can affect mortality.

2 What Is Affecting Weather-Related Mortality?

2.1 Meteorological Issues

The combination of temperature, wind, and humidity produce an "apparent temperature", which is the perceived temperature to the human body (Steadman 1984). Healthy persons have efficient heatregulatory mechanisms which cope with increases in apparent temperature up to the threshold condition. When exposed to heat, the body can increase radiant, convective, and evaporative heat loss by methods such as vasodilation (enlargement of blood vessels) and perspiration (Horowitz and Samueloff 1987; Diamond 1991). In addition, acclimatization to oppressive conditions can occur during several days with continuing exposure (Kilbourne 1992). However, there are obvious indications that the body cannot cope with oppressive conditions indefinitely. Total deaths increase substantially when temperatures exceed threshold values, especially for several consecutive days (Fig. 1; Kalkstein and Smoyer 1993).

Some recent studies have utilized a "synoptic climatological approach" to identify meteorological conditions which exceed the threshold. The synoptic approach identifies "air masses" which comprise the entire umbrella of air affecting individuals. Thus, the interaction of numerous meteorological factors is evaluated simultaneously, and their impact upon human health can be assessed (Kalkstein 1991a). For example, it is possible that a temperature of 35 °C with clear skies and high humidity may have a greater negative impact on human health than a temperature of 38 °C with cloudiness. Thus, evaluating temperature and humidity in the absence of other meteorological factors which might impact human well-being is less desirable. Each day is assigned to a particular air mass, allowing for the identification of "offensive air masses", which contain days with mortality that exceeds the baseline for a locale. For

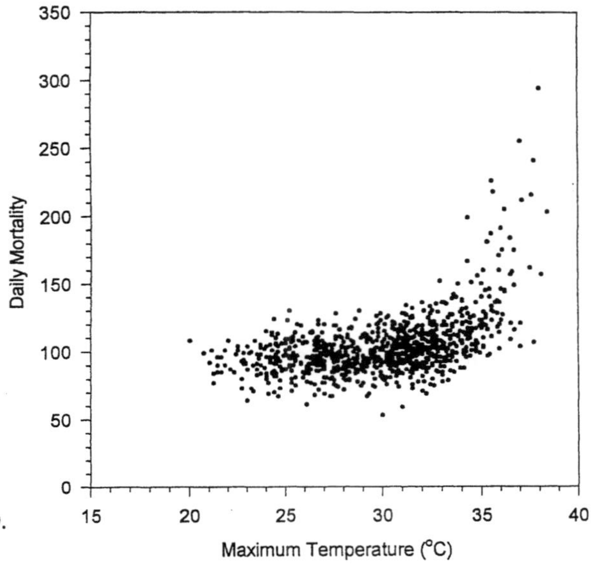

Fig. 1. Relationship between maximum temperature and mortality: Shanghai; 1980–1989. (Kalkstein and Smoyer 1993)

example, in Shanghai, an air mass which included days with high afternoon temperatures, low wind speeds, and low humidity was deemed offensive (Tan 1994). It is possible to determine which factors within the offensive air mass have the greatest negative impact upon human mortality. The synoptic procedure has been employed within the "weather/health watch-warning system" for Philadelphia, which is discussed later in this chapter.

The elderly and very young appear to be disproportionately affected by offensive air masses, as they possess a limited ability to adapt to stressful conditions. There is evidence that individuals living in poverty, as well as urban populations in developing countries such as Egypt and China, are particularly vulnerable to heat stress. Poor housing conditions, the exacerbation of stress because of the "urban heat island", and lack of access to air conditioning are frequently cited as primary causes (Kilbourne 1989; Mestel 1995). In addition, complete acclimitazation (the process by which a non-native can tolerate conditions as well as a native) may take up to several years (Babayev 1986), rendering individuals vulnerable to the vagaries of weather for a considerable period of time.

Although the most direct impact of heat stress on the human body is the onset of heat exhaustion or heatstroke, it appears that mortality increases associated with hot weather are related to a variety of causes. For example, deaths from cardiovascular, respiratory, and immune system disorders, as well as accidents, appear to increase during stressful weather conditions, and surprisingly, heat stroke and heat exhaustion represent only a small proportion of the mortality increase (Larsen 1990a, b; IPCC 1996). Thus, during hot weather, mortality from all causes may be more than double the long-term mean.

One of the major hazards of a global warming is the increase in frequency of the very stressful conditions which are most closely associated with a dramatic rise in mortality. For example, it has been estimated that the frequency of extremely hot days in Washington, DC, USA, and of extraordinarily hot summers in England, like that of 1976, could increase by one to two orders of magnitude by the year 2050 (Leaf 1989; Climate Change Review Group, UK 1991). During the 1980s in Missouri, USA, an increase of 2–3 °C in the average summer temperature was associated with a much larger number of days above the apparent temperature threshold, and a sevenfold increase in heat-related deaths resulted during that decade (Centers for Disease Control, CDC 1989).

Estimates of present-day heat-related mortality have been established for a number of locales in the developed and developing world. The most recent analyses of concurrent meteorological and mortality data in cities in the USA, Canada, the Netherlands, China, and the Middle East provide confirmatory evidence that overall death rates rise during heat waves (see next section; also Kalkstein and Smoyer 1993; Kunst et al. 1993). In addition, it is clear that medical examiners grossly underestimate the number of deaths attributed to heat due to constraints in defining a heat-related death (Kalkstein 1995). For this reason, some cities are now beginning to look at deaths above a mean mortality value to determine the magnitude of heat-related deaths. After the disastrous 1995 heat episode in Chicago, for example, the public health office in that city raised the mortality total attributed to heat from about 500 deaths to over 700 using this approach (Chicago Tribune 1995). Other city health departments are beginning to follow suit, and are becoming more aware that heat is probably the most important direct weather-induced cause of death.

The relationship between weather and cold-related mortality is less clear, and disagreements appear in the literature. In many countries, the death rate in winter is 10–25% higher than during the rest of the year. However, the existence of a temperature threshold is less evident in winter than in summer (Kalkstein and Davis 1989). Jendritzky (1995) found no threshold in Germany, and mortality increased linearly as cold stress increased (Fig. 2); this is confirmed in a recent study by Khaw (1995) . Although deaths from cardiovascular disease are important in both seasons, influenza, pneumonia, and increased mortality from a variety of accidents appear to be of greater significance in winter than in summer (Kunst et al. 1993; CDC 1994; Langford and Bentham 1995). For example, in four US cities (Birmingham, Cleveland, Philadelphia, and Seattle), mean mortality from cardiovascular disease increased by about 5–7% during winter (as compared to summer). Values for pneumonia and influenza increase by about 30%, and mortality from other respiratory infections by about 10% (NCHS 1978).

In some cold-weather locales, such as Montreal, Canada, and Minneapolis, USA, the increase in mortality with decreasing winter temperature is slight (Kalkstein 1988; Frost and Auliciems 1993). In less extreme climates, such as the Netherlands and Brisbane, Australia, mortality rates rise linearly with decreasing wintertime temperature (Keatinge et al 1989; Frost and Auliciems 1993;

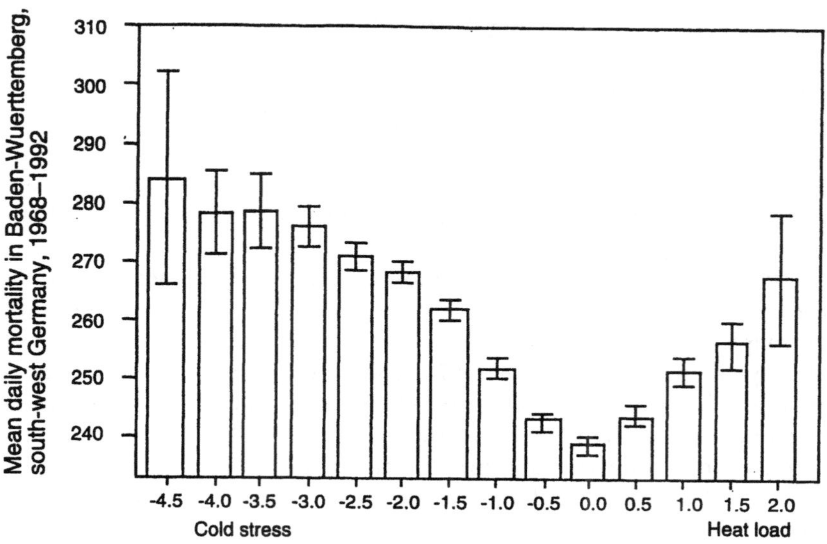

Fig. 2. Relationship between heat load and mortality in south-west Germany. (Jendritzky 1995)

Kunst et al. 1993). People in temperate environments may be less accustomed to extreme cold, while in colder climates, behavioral responses such as cold avoidance may be a dominant thermoregulatory process when very cold temperatures occur. From a physical standpoint, the increase in some diseases, such as cardiovascular, in winter may reflect an increased cold-induced tendency for the blood to clot, perhaps exacerbated by the fibrinogen-enhancing effect of winter respiratory infections (Keatinge et al. 1989; Woodhouse et al. 1994).

There is also some evidence that, in temperate climates, stormy, rather than very cold, weather is responsible for much of the increase in excess mortality in winter (Ragot and Padgett 1976; Glass and Zack 1979). Thus, in some places, heart attack related to overexertion may be a major cause of wintertime death (Faiche and Rose 1979). However, a greater propensity for respiratory infections may also be a result, as people are confined to poorly-ventilated places during inclement winter weather, increasing the probability of aerosol transmission (Curwen 1990; CDC 1994). The fact that the average lag time between stressful weather and subsequent mortality increase is much longer in winter than in summer supports this possibility (Kalkstein 1988; Larsen 1990a; Kunst et al. 1993). An unresolved question is whether a small rise in winter temperatures would encourage outdoor activities to such an extent that people would successfully avoid aerosol transmission. This question is central to the argument that a global warming would increase total weather-related mortality significantly, as summer increases would not be compensated by winter decreases. This is discussed later in this chapter.

2.2 Socioeconomic Issues Affecting Weather-Related Mortality

A number of nonmeteorological factors may influence weather-related mortality, especially in summer. Many relate to the modification of the environment which occurs with increasing urbanization. The following discussion describes the impact of three such factors: air conditioning, socioeconomic status, and the within-season variation in health impact (i.e., more people die early, rather than late, in the summer season).

In an attempt to determine if air conditioning is a mitigating factor in reducing heat-related mortality, a study was developed which compares mortality totals through a 25-year period when access to air conditioning increased dramatically in the US (US EPA 1997). The study evaluated mean annual summer mortality for a number of years between 1964 and 1988 for eight cities: Birmingham, Chicago, Memphis, New Orleans, New York, Philadelphia, Phoenix, and St. Louis. A determination of trends in annual mortality was made, with the assumption that a decreasing temporal trend may very well be attributed to increased human access to air conditioning.

There is no doubt that the number of homes with air conditioning increased rapidly during this period. In St. Louis, estimates indicate that 40% of homes had air conditioning in 1965 and this number had increased to 91% by 1992 (Stern et al. 1993). Similar increases are noted for other northern and midwestern cities (Fig. 3). However, determining whether decreases in annual mortality during this period are attributable to air conditioning accessibility is a more challenging task than simply evaluating interannual total mortality trends. There are other factors which could confound interannual mortality

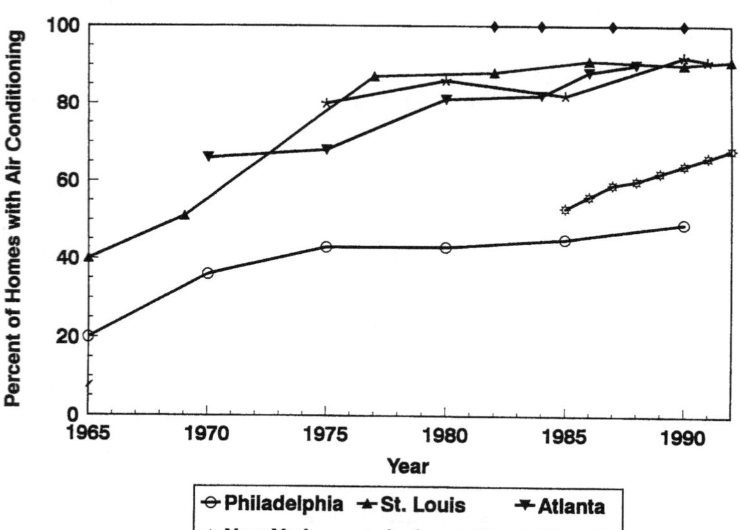

Fig. 3. Historical air conditioning saturation data. (Stern et al. 1993)

variability. The two most important are demographic changes during this period (if the population has increased, obviously there is the potential for mortality increases) and improvements in health care and medical facilities.

An attempt was made to remove these extraneous factors from the interannual trend. First, the mortality data were standardized for changes in population in each evaluated locale (Lilienfield and Lilienfield 1980; Kalkstein et al. 1997a). Second, in an attempt to account for improvements in health care, two interannual mortality trends were developed for each city; one for days within the offensive air mass and another all other days, when the weather was benign. It was assumed that the impact of improved health care should be a relative constant for both groups of days. However, virtually all heat-related deaths were found only during days within the offensive air mass, when air conditioning usage is greatest. Thus, the difference in slope between nonoffensive and offensive air mass days would be expected to provide an indication of air conditioning impact. If air conditioning does have a mitigating effect, the interannual mortality slope should be inverse and steeper for the days within the offensive air mass.

The results suggest that air conditioning has had some effect upon heat-related mortality. For a number of the evaluated cities, the interannual trend in mortality is decreasing (Fig. 4). This is especially the case for New York, Chicago, St. Louis, and Philadelphia, where standardized mortality during both offensive and nonoffensive air mass days show a general decline through the study period which is significant at the 0.10 level. In a number of cases, the slope is significant

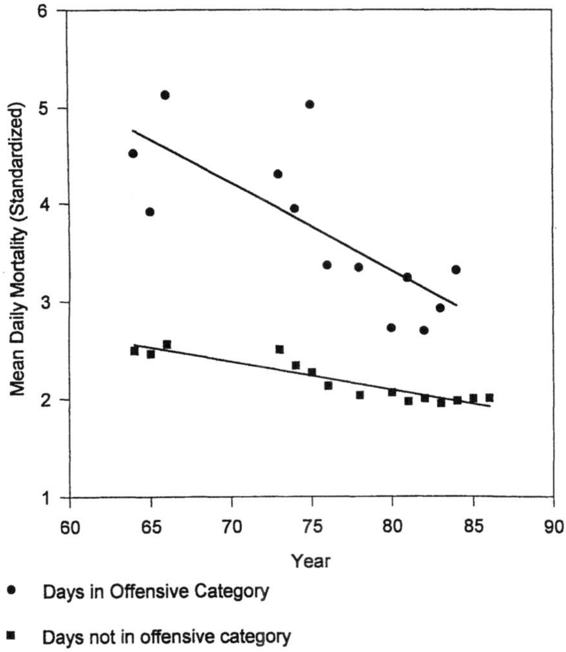

Fig. 4. Treads of mean daily mortality, New York, 1960–1990

at the 0.01 level. For the more northerly cities, the negative slope for stressful weather days is considerably greater than the slope for nonstressful days. These are cities where air conditioning availability has increased substantially during the period of record.
These results suggest the following:

- The negative slope of the interannual trend for the nonoffensive air mass days is primarily attributed to health care and public health improvements through the period.
- The steeper negative slope for the offensive air mass days is attributed to both health care improvements *and* the role of air conditioning in lessening heat-related deaths.
- The *differential* in slope between the offensive and nonoffensive air mass days is a measure of the impact of air conditioning in reducing heat-related days.

Results indicate that air conditioning has reduced heat-related mortality since 1964 by about 13% in New York, 11% in Philadelphia, and between approximately 10 and 20% for other cities in this study (Kalkstein 1997b). In addition, results indicate that the saturation of air conditioning will be almost 100% in these US cities by the year 2050 (Stern et al. 1993), and possibly as early as 2010 (GRI 1992). Thus, we are close to achieving the maximum mitigating impact of air conditioning.

At first glance, it is surprising that air conditioning does not have a greater impact in lessening heat-related mortality. However, many of the individuals most susceptible to heat-related illness are the elderly and poor, many of whom still do not have direct access to air conditioning. In addition, it is noteworthy that *many* causes of deaths, including those that are not related to direct exposure to heat (e.g., cardiac problems, stroke), increase during stressful weather. Thus, it is possible that some people do not require direct exposure to hostile external environments to be negatively affected by these environments (Kalkstein and Davis, 1989). The knowledge that unpleasant external conditions exist might be sufficient to contribute to negative reactions. A study measuring recovery speed of surgery patients and window views tends to support this contention. Twenty-three surgical patients assigned to hospital rooms with windows looking out on a pleasant landscape had shorter postoperative hospital stays than 23 matched patients in similar rooms but with windows facing a brick wall (Ulrich 1984). Thus, perception of hostile environments, even if exposure is not direct, may have a significant impact on weather-related mortality.

Beyond the availability of air conitioning, other social factors, such as poverty, have an impact on weather-related mortality. The determination of the impact of poverty on mortality was evaluated in four US cities (New York, New Orleans, Atlanta, and St. Louis; Smoyer 1993). Although death records do not indicate the socioeconomic status of the deceased, information such as age and ethnicity was used in this study to identify those in the population who were poor. The results suggest that poverty renders segments of the population more sensitive to heat-related mortality. Poverty appears to be an especially strong factor in

St. Louis, and to a lesser degree in New Orleans; however, no such association was found in Atlanta.

A particularly interesting response was uncovered in New York City, where both poor and nonpoor populations appear highly susceptible to heat-related mortality within the city. In addition, poor and nonpoor groups appear not to be susceptible in the surrounding counties. This result suggests that in New York the "urban heat island" impact is considerably stronger than any socioeconomic factor in affecting heat-related mortality. Both poor and nonpoor within the city are affected adversely by the heat, and considering their disparate abilities to cope with this condition (e.g., availability of air conditioning, housing quality differences), it appears that socioeconomic factors have a limited impact.

The Smoyer (1993) study also detected an age-related factor, and poor elderly people, especially in St. Louis and New York, seem particularly vulnerable to heat-related mortality. In addition, those areas with the highest population density, lowest percentage of detached housing, and highest percentage of housing constructed before 1939 contained the most susceptible populations (Smoyer 1993).

In a discussion paper by Chestnut et al. (1993a), an attempt was made to correlate mean seasonal heat-related mortality (for the period 1964–1980) with various socioeconomic characteristics for 15 US cities. Some of the results accord with the Smoyer evaluation; for example, statistically significant socioeconomic characteristics included the percentage of the population over age 65 and the percentage of homes built before 1950. However, several additional characteristics proved to be important. Cities with a larger percentage of labor in manufacturing demonstrted higher heat-related mortality totals. In addition, cities with a higher proportion of houses with incomplete plumbing, with high central city population densities, and with low percentages of population with at least a high school education also exhibited increased sensitivity to heat-related mortality. It is certainly possible that these socioeconomic characteristics are not causal mechanisms but are instead coincidentally associated with older eastern and midwestern cities whose climatic characteristics render them more vulnerable to high heat-related mortality totals. Thus, Chestnut et al. (1993) readily acknowledge that a more detailed study is warranted.

One question which arises from heat-stress mortality analysis is whether people who have died during heatwaves would have died soon afterward even if no stressful weather had occurred (mortality displacement). A recent study has addressed this question by evaluating mortality during and shortly after extreme heat events for three US cities: Birmingham, New York, and St. Louis (US EPA 1997). Several episodes of high mortality for each city which occurred during extreme heat events were selected for analysis.

The evaluation is based on the premise that, if mortality displacement is a significant factor, daily mortality totals shortly after a severe heat episode would be less than the long-term mean. Figure 5 illustrates that the standardized total mortality during and after a New York heat wave in 1966 was well above the long-term mean. On one day the mortality rate was almost three times the mean

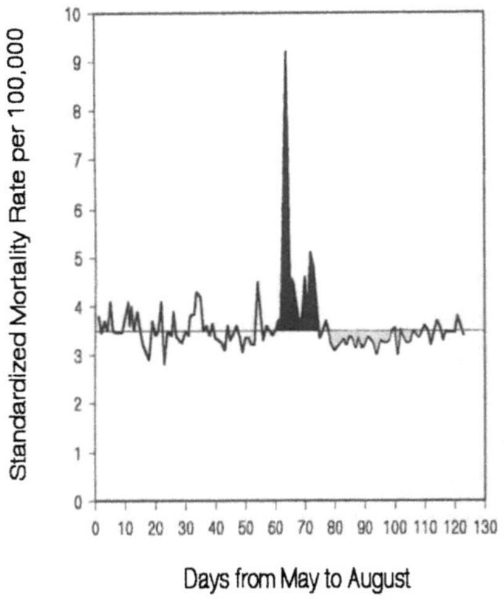

Fig. 5. Daily summer mortality around a New York heat wave, 1966. *Solid shading* shows mortality during the heat wave; *stippling* shows mortality in the immediately following period

value. However, during the period immediately after the heat episode, it was apparent that daily mortality was somewhat below the mean until mid-August. It was assumed that this lower death rate is due to mortality displacement.

A quantitative method was devised to estimate the proportion of mortality due to displacement. For example, during the 1966 heat episode in New York, a total of 19.8 deaths per 100 000 were accumulated above the long-term mean for the period of the heat wave. For the period which was significantly lower than the long-term mean after the heat episode (presumably due to mortality displacement), a total of 8.6 deaths per 100 000 were accumulated below the long-term mean. This suggests that about 43% of the total extra deaths during the heatwave was due to displacement.

Some degree of displacement was found for all evaluated heat episodes (Table 1). The intracity values were fairly consistent (the three episodes evaluated for New York demonstrated a displacement value between 38.7 and 43.3%), although significant intercity differentials were uncovered (the Birmingham displacement value for the 1980 heat episode was 72%, and displacement values for St. Louis were considerably lower than New York's). Thus, it appears that a significant proportion of the excess mortality which occurs during a heat episode would have occurred nevertheless shortly afterward. However, it is equally important to note that in New York, and especially St. Louis, the majority of the deaths occurring during a heat episode represent mortality which would not have occurred shortly afterward if a heat episode were not present. It appears that only one in five excess heat-related deaths in St. Louis is a result of displacement.

A similar study using a different approach was recently completed and uncovered mortality displacement proportions roughly similar to those cited in the study above (Kalkstein et al. 1997b). Displacement values for New York were

Table 1. Percentage of mortality "displacement" in six heat episodes in the USA

Station, year	Mean daily summer mortality (rate per 100 000)	Sum above mean[a]	Sum below mean[b]	Ratio[c]
New York, 1966	3.50	19.757	8.559	.4330
New York, 1975	3.02	8.651	3.446	.4025
New York, 1984	2.67	7.754	3.004	.3870
St. Louis, 1966	2.90	33.365	7.344	.2200
St. Louis, 1980	2.82	17.130	2.665	.1556
Birmingham, 1980	2.44	14.160	10.19	.7200

[a] Sum of all mortality during heat episode (see Fig. 5 for calculation).
[b] Sum of all mortality after heat episode and during period when mortality was statistically significantly lower than preheat episode levels (see Fig. 5).
[c] Sum below mean/sum above mean.

estimated at 28% and those for Philadelphia at 29%; the average US figure was 26%.

Considering that mortality displacement is a significant feature during many heat episodes, it will be necessary to adjust all retrospective and prospective estimates of heat-related mortality accordingly.

3 Weather-Related Mortality in a Warmer World

The impact of a global warming on weather-related mortality is not easy to ascertain, as it is difficult to assess a population's ability to adapt to a gradual climate change (Kalkstein 1995). However, considering that death rates rise sharply during hot weather, and considering that the frequency and intensity of such events will increase according to virtually all climate prediction models, it is likely that heat-related deaths and illnesses will increase substantially (Haines et al. 1993; WHO/WMO/UNEP 1996). Recent estimates indicate that in very large cities, such as Shanghai and New York, climate change will cause several thousand extra heat-related deaths annually by the middle of the twenty-first century (Table 2). The impacts would be greatest in mid-latitude cities with infrequent, but extreme, heatwaves. These are the areas which exhibit the greatest number of heat-related deaths today, and include large cities in eastern and midwestern North America, western Europe, northern India, and China. Subtropical and tropical cities with higher mean summer temperatures seem less vulnerable, due in part to socioeconomic factors such as more heat-efficient housing, and, in some cases, greater access to air conditioning (Rogot et al. 1992; Kalkstein and Smoyer 1993). Although there may be some adjustments made by the population through behavioral acclimatization, it is unlikely that urban housing in these mid-latitude cities will change in response to global warming. For example, the tenements in poor neighborhoods of St. Louis and New York,

Table 2. Present-day figures and future estimates of total summer deaths[a] in selected cities

Location	Present mortality	GFDL 90 Year 2020 Unacc.	Acc.	UKTR Year 2050 Unacc.	Acc.	Year 2020 Unacc.	Acc.	Year 2050 Unacc.	Acc.
United States									
Atlanta	78	191	96	293	147	247	124	436	218
Dallas	19	35	28	782	618	1364	1077	1360	1074
Detroit	118	264	131	419	209	923	460	1099	547
Los Angeles	84	205	102	350	174	571	284	728	363
New York	320	356	190	879	494	1098	683	1754	971
San Francisco	27	49	40	104	85	57	47	76	62
Canada									
Montreal	69	121	61	245	124	460	233	725	368
Toronto	19	36	0	86	1	289	3	563	7
People's Republic of China									
Shanghai	418	1104	N/A	2950	N/A	1308	N/A	1486	N/A
Egypt									
Cairo	281	476	N/A	830	N/A	839	N/A	1024	N/A

GFDL 89, Geophysical Fluid Dynamics Laboratory; *UKTR*, United Kingdom Meteorological Office General Circulation Model; *Unacc*, Unacclimatized population; *acc*, acclimatized population.

[a] Numbers represent average summer-season heat-related deaths for each city under each climate change scenario. For example, during a typical summer today, 78 extra deaths occur in Atlanta from heat-related causes and, assuming no acclimatization, this number rises to 191 deaths. Numbers assume no change in populations size and age distribution.

which are especially unsuited to hot conditions because of their black roofs, red brick exteriors, and lack of ventilation, will not be demolished and replaced by more heat-efficient housing if the climate warms. At best, only partial acclimatization can be expected if global warming occurs, and mortality and heat-related illnesses are likely to increase, especially from cardiorespiratory causes (IPCC 1996).

The impact of warmer winters on cold weather mortality is more difficult to determine. There are some who believe that, in a warmer world, cold-related mortality will decrease sharply. Some have stated that this decrease could substantially offset the predicted summer increase (Rango 1984; Tan 1994). One recent British study has forecast that approximately 9000 fewer winter-related deaths would occur annually by the year 2050 in England and Wales under typical climate change scenarios that entail 2–2.5 °C wintertime increases (Langford and Bentham 1995). About half of those avoided deaths would have been from ischemic heart disease and stroke, with chronic bronchitis and pneumonia each contributing 5–10%. Conversely, there are other studies stating that winter mortality reductions will be considerably less than summer increases if a global warming occurs (Kalkstein 1993; US EPA 1997). These analyses contend that increased respiratory infections in winter are a primary reason why winter mortality is higher than summer. Since such infections depend on aerosol transmission in poorly ventilated places, it is important to determine if a small rise in winter temperatures will encourage outdoor activities to a level where risk is reduced. Considering that annual influenza outbreaks do not appear to correlate with mean winter or mean monthly temperature (CDC 1994), this may indicate that the predicted temperature increases attributed to global warming will not drive people outdoors during winter. Thus, droplet-transmitted diseases and deaths may not be reduced substantially if the climate warms as predicted. However, other studies contend that even a modest temperature rise in winter would encourage outdoor activities to such an extent that people would more successfully avoid aerosol transmission (Langford and Bentham 1995). If this were true, however, it would be expected that epidemics of droplet-transmitted diseases would be most likely during the colder winters, and historically, this has not been the case.

In summary, the general consensus points to a sizable increase in heat-related mortality, and a more uncertain but probable decrease in cold-related mortality if the earth warms as expected. It is likely that the heat-related increases will be considerably greater than the cold-related decreases, leading to a sizable net increase in weather-related mortality under global warming conditions.

4 Mitigating Measures to Lessen the Impact of Weather-Related Mortality

A number of strategies are possible to reduce the impact of weather on present and future human mortality. For example, the introduction of protective technologies (e.g., better insulation of buildings, extension of air conditioning access to developing country populations) would be helpful (WHO/WMO/UNEP 1996). In addition, there are a number of cost-effective options to lower summer temperatures in dwellings while saving billions of dollars in energy costs (Mestel 1995). Widespread use of white paint and light-colored roofing materials might decrease in-home temperatures sufficiently to lessen heat-related mortality significantly.

A pressing requirement is for an adequate surveillance system to alert the public and city agencies responsible for public health that dangerous weather is predicted. Such a system was in place in Philadelphia during the summers of 1995–1997, and it seems to have improved interventions by that city's Department of Public Health (Kalkstein 1995). Some of the intervention techniques used by Philadelphia include:

- Institution of the "buddy system". Volunteers within the neighborhood go door-to-door to make certain the elderly and infirm are fine.

- Activation of the "hot line". The Philadelphia Department of Public Health activates this telephone line when a warning is issued, so people can call and be given advice on how to cope with the heat. The hot line may also be used for emergencies for people who are under severe heat stress.
- Establishment of a "proactive team". This group of Department of Public Health workers and nurses travels to nursing homes and other facilities before problems are reported to make certain that proper procedures are being followed during the heat wave.
- Establishment of a "reactive team". These represent injury prevention workers who respond to hot line queries and go directly to homes where people are in distress. They assist such people and teach them heat stress avoidance procedures.
- Cooperation with the Philadelphia Corporation of Aging. This agency answers all hot line calls.
- Cooperation with local utility companies. The Department of Health alerts the utility companies that a heat-health warning has been called. No electrical service can be terminated to individuals when a warning is in effect.

The Philadelphia Watch/Warning System is based upon synoptic climatological methodology, and it not only identifies offensive air masses, it predicts their arrival up to 48 h in advance, permitting the city to take the steps described above. The system distinguishes days within the air mass which are most likely to have increased mortality (Fig. 6). A "heat-health watch" is issued by the Philadelphia Commissioner of Health if an offensive air mass is forecast within 48 h, an "alert" is issued if it is forecast within 24 h, and a "warning" (requiring National Weather Service concurrence) is issued if the air mass is imminent and extra heat-related deaths are predicted. Early evaluations suggest that this system is more effective than previous approaches, such as those based on the National Weather Service "heat index". The heat index is a measure of apparent or perceived temperature, and a forecast of heat indices above 105 °F (40.6 °C) for three consecitive hours is necessary for the National Weather Service to issue a heat advisory. The arbitrary nature of this heat index threshold, and the fact that it is not based on direct human helath responses to weather, render it inferior to the synoptic-based system presently being utilized by Philadelphia.

Improved coordination on a local, regional, and global level is needed if heat-related mortality is to be lowered. National standards to identify heat-related mortality, and large-scale warning systems are imperative. At present, the US EPA (Environmental Protection Agency) and National Oceanic and Atmospheric Administration are coordinating an effort to develop a national watch-warning system similar to the Philadelphia scheme. This is a timely effort, as excessive heat probably contributes to more mortality than any other direct weather-related cause.

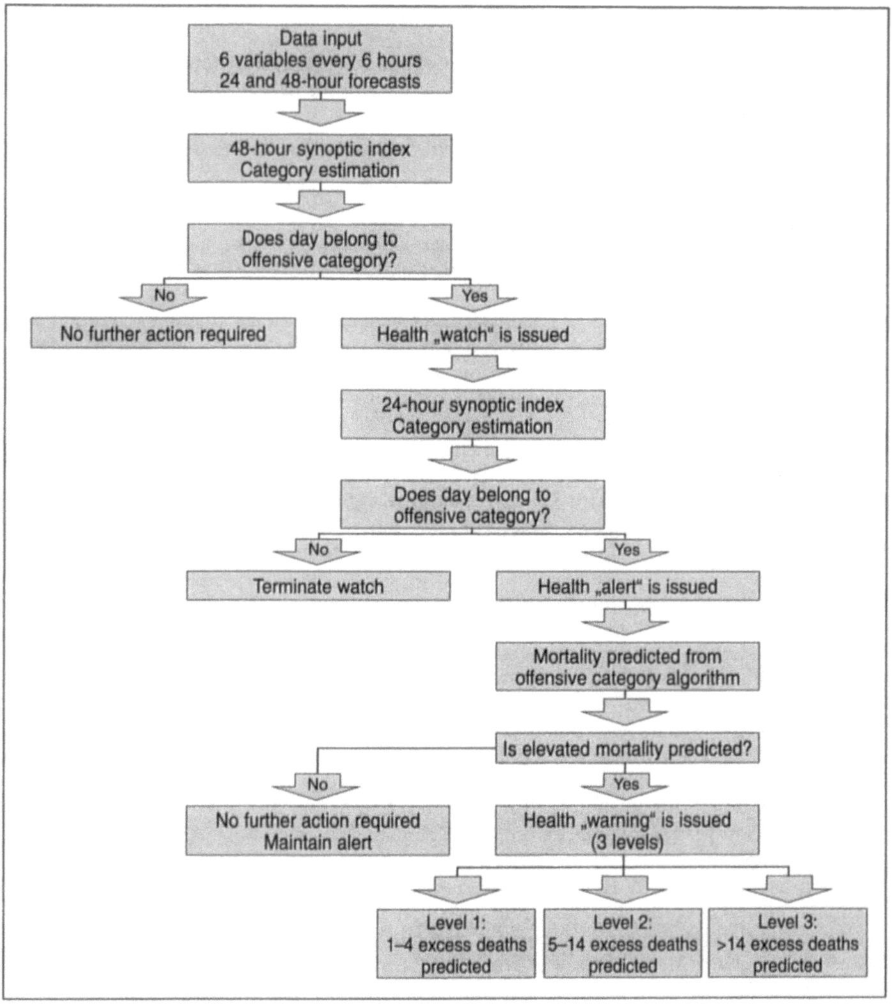

Fig. 6. The Philadelphia heat-stress watch/warning system

References

Babayev AG (1986) Some aspects of man's acclimatization to hot climates. In: Climate and Human Health. WHO/UNEP/WMO Int Symp Vol 2, Leningrad, pp 123

CDC (1989) Heat-related deaths – Missouri, 1979–1988. Morbid Mortal Week Rep 38:437–439

CDC (1994) Update influenza activity – United States and worldwide, 1993-93 season and composition of the 1994-95 influenza vaccine. Morbid Mortal Week Rep 43(10):179–183

Chestnut L, Breffle W, Castille E, Smith J (1993) Differences in summer mortality due to climate in USA cities. Report to the Climate Change Division, US EPA

Chicago Tribune (1995) Heat's toll worse than believed. September 21, 1995

Climate Change Impacts Review Group UK (1991) The potential effects of climate change in the United Kingdom. HMSO, London

Curwen M (1990) Excess winter mortality: a British phenomenon? Health Trends 1990, 22:169–175
Diamond J (1991) The rise and fall of the third chimpanzee. Radius, London
Ellis FP, Nelson F (1978) Mortality in the elderly in a heatwave in New York City, August, 1975. Environ Res 15:504–512
Faiche G, Rose R (1979) Blizzard morbidity and mortality: Rhode Island, 1978. Am J Public Health 69:1050–1052
Frost DJ, Auliciems A (1993) Myocardial infarct death, the population at risk and temperature habituation. Int J Biometeorol 37:46–51
Gas Research Institute (1992) 1992 baseline projections. GRI, Chicago
Glass RT, Zack MMJ (1979)Increase in deaths from ischaemic heart disease after blizzards. The Lancet 1:485–487
Haines A, Epstein PR, McMichael AJ, Kalkstein LS (on behalf of an international panel) (1993) Global health watch: monitoring impacts of environmental change. Lancet 342:1464–1469
Horowitz M, Samueloff S (1987) Circulation under extreme heat load. In: Dejours P (ed) Comparative physiology of environmental adaptations Vol 2. Karger, Basel, pp 94–106
IPCC (Intergovernmental Panel on Climate Change) 1996) Climate change 1995: impacts, adaptations and mitigation of climate change: scientific-technical analysis. Watson RT, Zinyowera MC, Moss RH (eds) Cambridge University Press, New York, 878 pp
Jendritzky G (1995) Umweltfaktor Klima. In: Beyer A, Eis D (eds) Praktische Umweltmedizin. Springer, Berlin Heidelberg New York
Kalkstein LS (1988) The impacts of predicted climate change on human mortality. Publ Climatol 41:1–127
Kalkstein LS (1991a) A new approach to evaluate the impact of climate upon human mortality. Environ Health Perspect 96:145–150
Kalkstein LS (1991b) The potential impact of global warming: climate change and human mortality. In: Wyman RL (ed) Global warming and human mortality. Routledge, Chapman and Hall, New York, pp 216–223
Kalkstein LS (1993) Health and climate change – direct impacts in cities. Lancet 342:1397–1399
Kalkstein LS (1995) Lessons from a very hot summer. Lancet 346(8979):857–859
Kalkstein LS, Davis RE (1989) Weather and human mortality: an evaluation of demographic and inter-regional responses in the United States. Ann Assoc Am Geogr 79:44–64
Kalkstein LS, Smoyer KE (1993) The impact of climate change on human health: some international implications. Experientia 49:469–479
Kalkstein LS, Barthel CD, Ye H, Smoyer K, Cheng S, Greene JS, Nichols MC (1997a) The differential impacts of weather and pollution on human mortality. EPA Office of Policy, Planning and Evaluation Monograph, document number EPA 230-R-94-019, US EPA Office of Research and Development, Washington, DC (in press)
Kalkstein LS, Tan G, Cheng S (1997b) The impact of mortality displacement and air conditioning on heat-related mortality: present and future implications, Report to US EPA Climate Change Division, Washington, DC, 24 pp (in press)
Keatinge WR, Coleshaw SRK, Holmes J (1989) Changes in seasonal mortalities with improvement in home heating in England and Wales from 1964–1984. Int J Biometeorol 33:71–76
Khaw KT (1995) Temperature and cardicovascular mortality. Lancet 345:337–338
Kilbourne EM (1989) Heatwaves. In: Gregg MB (ed) The public health consequences of disasters 1989. US Department of Health and Human Services. Atlanta, pp 51–61
Kilbourne EM (1992) Illness due to thermal extremes. In: Last JM, Wallace RB (eds) Public health and preventive medicine, Omit, 13th edn. Appleton, Norwalk, Lange, pp 491–501
Kunst AE, Looman CWN, Mackenbach JP (1993) Outdoor air temperature and mortality in the Netherlands: a time-series analysis. Am J Epidemiol 137:331–341
Langford IH, Bentham G (1995) The potential effects of climate change on winter mortality in England and Wales. Int G Biometeorol 38:141–147
Larson U (1990a) The effects of monthly temperature fluctuations on mortality in the United States from 1921 to 1985. Int J Biometeorol 34:136–145

Larson U (1990b) Short-term fluctuations in death by cause, temperature, and income in the United States 1930 to 1985. Soc Biol 37(3-4):172–187

Leaf A (1989) Potential health effects of global climatic and environmental changes. N Engl J Med 321:1577–1583

Lilienfield AM, Lilienfield DE (1980) Foundations of epidemiology. Oxford University Press, New York

McMichael AJ (1993) Global environmental change and human population health: a conceptual and scientific challenge for epidemiology. Int J Epidemiol 22(1):1–8

Mestel R (1995) White paint. New Sci March 25, 1995:34–37

National Center for Health Statistics (1978) Standardized micro-data tape transcripts. Department of Health, Education and Welfare, Washington, DC

Rango NR (1984) Exposured-related hypothermia mortality in the United States, 1970–1979. Am J Public Health 74:1159–1160

Rogot E, Padgett SJ (1976) Associations of coronary and stroke mortality with temperature and snowfall in selected areas of the United States. 1962–1966. Am J Epidemiol 103:565–575

Rogot E, Sorlie PD, Backland E (1992) Air-conditioning and mortality in hot weather. J Epidemiol 136:106–116

Smith HB, Tirpak DA (eds) (1989) The potential effects of global climate change on the United States. Report to Congress. EPA 230-05-89-050. US Environmental Protection Agency, Washington, DC

Smoyer KE (1993) Socio-demographic implications in summer weather-related mortality. Master's Thesis, University of Delaware, Delaware

Steadman RC (1984) A universal scale of apparent temperature. J Clim Appl Meteorol 23:1674–1687

Stern F, Neal B, Smith J (1993) Air-conditioning saturation in selected major urban areas. A report to the Climate Change Division, EPA. RCG/Hagler, Bailly, Boulder, Colorado, 10 pp

Tan G (1994) Potential effects of global warming upon human mortality in Shanghai and Guangzhou, China. Acta Sci Circumstant 143:68–373

Ulrich RS (1984) View through a window may influence recovery from surgery. Science 224:420–421

United States Environmental Protection Agency (1997) Pleminiary assessment of the benefits to the US of avoiding, or adapting to climate change. EPA Climate Change Division, Washington, DC (in press)

WHO/WMO/UNEP 1996) Climate change and human health. (McMichael AJ, Haines A, Slooff R, Kovats S (eds) Geneva, 297 pp

Woodhouse PR, Khaw KT, Plummer M, Foley A, Meade TW (1994) Seasonal variation of plasma fibrinogen and factor VII activity in the elderly: winter infections and death from cardiovascular disease. Lancet 343:435–439

Subject Index

MIX
Papier aus verantwortungsvollen Quellen
Paper from responsible sources
FSC® C105338
FSC
www.fsc.org

If you have any concerns about our products,
you can contact us on
ProductSafety@springernature.com

In case Publisher is established outside the EU,
the EU authorized representative is:
Springer Nature Customer Service Center GmbH
Europaplatz 3, 69115 Heidelberg, Germany

Printed by Libri Plureos GmbH
in Hamburg, Germany